MW00977748

Green Politics Is Eutopian

Green Politics Is Eutopian

Paul Gilk

WIPF & STOCK · Eugene, Oregon

GREEN POLITICS IS EUTOPIAN

www.wipfandstock.com

ISBN 13: 978-1-55635-776-3

Manufactured in the U.S.A.

The essay " A Green Critique of Socialist Agriculture" was originally published by Oberlin College in *Alternatives!* under the title "Preliminary Thoughts Toward an Understanding of Socialist Agriculture." "The Conscious Id" was also published in *Alternatives!*

The essay "The Meaning of Green Agriculture" appeared first in *Synthesis/ Regeneration 28* (Spring, 2002).

All three essays are reprinted here with permission.

We need some notion of what moved people to push steadily in one direction, for example towards the greater application of technology to production, or towards greater concentration of population. But what are often invoked are motivations that are non-moral.

—Charles Taylor,
The Ethics of Authenticity, p. 20

The West has no meaning in itself because the only value recognized by the theory of civilization is the refinement which is believed to increase steadily as one moves from primitive simplicity and coarseness toward the complexity and polish of urban life.

—Henry Nash Smith,
Virgin Land: The American West as Symbol and Myth, p. 267

Usually we use the term "civilization" for anything that is good about our humanity—for example, poetry and drama, music and dance, art and architecture, image and narrative. Correspondingly, to call individuals or groups, places or actions, "uncivilized" is normally a calculated insult. So I need to explain very clearly what I mean in this book by the "brutal normalcy of civilization." The point I wish to emphasize is that imperialism is not just a here-and-there, now-and-then, sporadic event in human history, but that civilization itself, as I am using that term, has always been imperial—that is, empire *is* the normalcy of civilization's violence.

—John Dominic Crossan,
God and Empire: Jesus Against Rome, Then and Now, p. 30

So what would a world look like that had viable local peace cultures on every continent? We can't work for something we can't imagine! We urgently need, individually, in our families, in our meetings, and in all the groups we work with, to spend significant periods of time in deep reflection about and envisioning of an earth-world that has become the peaceable garden it was created to be. A more local earth-world, in which all living things are attuned to one another and learn from one another. A world full of music, the joy of work, and the joy of play. Our vision will empower our action as each of us begins to use the tools we have, in the settings in which we move, in ways that will sustain the peaceable garden. We are all gardeners, and the vision is the journey.

—Elise Boulding,
"Peace Culture: The Vision and the Journey"
in *Friends Journal* (September 2005), p. 16

The study of history always implies a study of its alternative.

—James Carroll,
Constantine's Sword: The Church and the Jews, p. 17

For my wife Susanna

Contents

Acknowledgments

There are several people who have encouraged or aided the intellectual journey taken in these essays. Maynard Kaufman, my dear Mennonite professor/farmer friend from southwest Michigan, heads the list. Next comes Dennis Boyer, Vietnam veteran and labor lawyer, who founded a small political party that transformed itself into the Wisconsin Greens. Then came an affiliation with (the now-defunct) *North Country Anvil* magazine, a regional Minnesota quarterly, an affiliation that introduced me to Jack Miller, Pauline Redmond, and Rhoda Gilman. Friendship with these people in particular kept me intellectually alive over most of the past thirty years. My life would have been significantly different without them. I thank them one and all.

Jennifer McEwen put my ragged manuscript on disk. Mary Mangold and Carol Ann Okite enabled me (a willful computer idiot) to get that disk ready for publication. Lynn Kordus did a brisk and competent copyedit. Finally, it was Mike Miles, congressional Green candidate, catholic-worker farmer and perennial peace activist, who directed this manuscript to its present publishing home. I hope these essays are worthy of their efforts.

Introduction

HERE ARE ESSAYS, WRITTEN over nearly a twenty-five year span, that attempt to explore the deeper, buried dimensions of Green politics, or, more importantly, Green culture.

If we ask a simple question—namely, what is it about the modern world that has given rise to the international Green phenomenon—we are led not only to a contemporary environmental critique, but to a quest for the roots of those forces that constitute the ecological assault. These essays consist of such an exploration.

I make no claim to be comprehensive, much less exhaustive, in this analysis. These are Green forays into difficult, but critically important, terrain. These forays include the meaning of utopia, the conventional conception of socialist agriculture, the dynamics of "backwardness" versus "progress," our saturation with the prevailing secular religion called Civilization—and a ragged, organic bouquet of other relevant topics.

My core conviction can be stated quite simply: such apocalyptic global realities as weapons of mass destruction and global warming/climate change tell us that we face transformation or disaster—either caring and sharing or hatred and destruction. The accrued lethality of the (largely male) enterprise of civilization, both economically and militarily, now threatens all mammalian life on Earth. This is not hyperbole. This is not hysterical exaggeration. This is the simple and terrifying truth.

In my estimation, there are only two "tools" that point, not merely toward survival, but toward a restored Earth with a humane culture and ecological economy, such as can be achieved, given extinctions, climate change, and accrued toxicity. The first tool is the *ethical* core of all true spiritual traditions: compassion, forgiveness, sharing, moderation, simplicity, modesty, selflessness, and love. The second tool is the slow, somewhat bumbling, but steady congealing of the Green political vision, a vision that, of necessity, engages in politics, but has its heart and soul invested in the yearning for and creation of Green culture.

These essays are not a how-to manual for being political. They won't tell you how to organize a local Green chapter or how to get Green candidates elected to your city council or county board. Before many Greens *do* get elected, we had better come to grips with some potent underlying issues or Green politics will be just another drifting ship without a rudder. A major change in the direction of Green culture requires a strong political will, and that will had better be deeply immersed in transformative spiritual ethics.

While it is our obligation and our responsibility to hope for an elegant, ecological future, such hope requires not only committed action, but also deeply ethical understanding. I pray that in these essays I am leading no one in a false direction.

1

E. F. Schumacher

Utopian or Eutopian?

In England and America there is probably no single person who has influenced the popular social movements of ecological preservation and environmental protection, organic agriculture, appropriate technology, and rural regeneration so much as the late E. F. Schumacher.

It is no coincidence that Theodore Roszak, in his Introduction to Schumacher's *Small is Beautiful*, links Schumacher's thought to "the tradition we might call anarchism, if we mean by that much-abused word a libertarian political economy that distinguishes itself from orthodox socialism and capitalism by insisting that the *scale* of organization must be treated as an independent and primary problem."[1]

The anarchist designation becomes even more interesting when we realize the thinkers with whom Roszak links Schumacher—Peter Kropotkin, Leo Tolstoy, William Morris, Mohandas Gandhi, Lewis Mumford, Paul Goodman, Murray Bookchin—have also been called utopians. Furthermore, since political philosophy places anarchism (at least its communal expression) under the general heading of "utopian socialism," we must therefore deal with the concept of utopia.

There are two distinct forms of "utopian" thought, as Lewis Mumford demonstrates in *The Story of Utopias*: ". . . for, as Professor Patrick Geddes points out, Sir Thomas More was an inveterate punster, and Utopia is a mockname for either Outopia, which means no-place, or Eutopia—the good place."[2] So we have "no-place" versus "the good place." "No-place" has often been depicted as a refined urban system: the ideal, the perfect, and the permanent. "The good place" has received considerably less attention than "no-place," but it can be characterized as the real, the whole, and the stable.

Despite the confusing contradictions in the respective titles, we can take two late-nineteenth-century novels as clear examples of the "no-place"/"good-place" division: Edward Bellamy's *Looking Backward* and William Morris's *News from Nowhere*. The contradiction is clarified by noticing that Bellamy's "ideal" story is set entirely in a city, while Morris's "real" tale is situated in the countryside. Bellamy's story is of an authoritarian, if also benevolent, urban hierarchy that directs a city-as-machine, while Morris's tale is of robust community-oriented physical life in a classless and unspoiled countryside.

So if anarchism distinguishes itself from socialism and capitalism by its emphasis on organic scale, then eutopian thought distinguishes itself from mainline utopian thought by its focus on natural life and the culture of the countryside. Now if mainline utopian thought is abstract and systems-oriented—concerned with the perfect, the permanent, and the ideal—what are we to make of Schumacher's well-known goals of health, beauty, and *permanence*? Perhaps we need to look more closely into the distinctions between permanence and *stability*. Even healthy organisms die through organismic deterioration. Natural beauty rests in part on healthy organisms that age and die. Health, to employ a whimsical word, is biodegradable, and so is beauty, for it also finds its expression in a constant process of creation and decay. The flux of creation and decay is the balancing process in the wheel of life—a process that underlies the patterns of both health and beauty, and of life itself.

If we can agree that health and beauty are based on organic fluctuation, then what are we to make of Schumacher's third ideal, permanence? The word in *Webster's* derives from the Latin *permanens* or *permanere*, to stay to the end. Additional meanings are: continuing or enduring in the same state, place, without marked change; not subject to alteration; lasting or abiding. Given the general tone of these definitions, permanence seems to be an ideal of considerably different quality than either health or beauty. Where both health and beauty must adapt to the flow of time and circumstance, permanence suggests a goal-oriented inflexibility, a desire for immortality. Neither health nor beauty endures in the same condition without change; on the contrary, their emergence is dependent on previous decay, and their own passing is assured.

It is, of course, true that nothing in our natural world is permanent in any absolute sense—not even radioactive waste. So we have to investigate Schumacher's puzzling inclusion of permanence into the company of health and beauty by asking what *aspires* to permanence. What aspires to permanence is a specific kind of human identity, the most objective illustration of which is civilization: that unprecedented slave-work system generated through the city and seeking to impose a permanent, goal-oriented, linear conception of time on what were (and always had been) nonurban cultures of natural connectedness and cyclical renewal. There is linkage here between permanence, civilization, and utopia.

In an essay entitled "Utopia, the City, and the Machine," included in Part Three of *Interpretations and Forecasts: 1922–1972,* Lewis Mumford says that "utopias from Plato to Bellamy have been visualized largely in terms of the city," and that the "concept of utopia" is a "derivation from an historic event: that indeed the first utopia was the city itself."[3] In an analysis of Plato's Second Book in the *Republic*, Mumford points out that:

> . . . Plato came near to describing the normative society of Hesiod's Golden Age: essentially the pre-urban community of the Neolithic cultivator, in which even the wolf and the lion, as the Sumerian poem puts it, were not dangerous, and all the members of the community shared in its goods and its gods—in which there was no ruling class to exploit the villagers, no compulsion to work for a surplus the local community was not allowed to consume, no taste for idle luxury, no jealous claim to private property, no exorbitant desire for power, no institutional war. Though scholars have long contemptuously dismissed the 'myth of the Golden Age,' it is their scholarship, rather than the myth, that must now be questioned.
>
> Such a society had indeed come into existence at the end of the last Ice Age, if not before, when the long process of domestication had come to a head in the establishment of small, stable communities, with an abundant and varied food supply: communities whose capacity to produce a surplus of storable grain gave security and adequate nurture to the young. This rise in vitality was enhanced by vivid biological insight and intensified sexual activities, to which the multiplication of erotic symbols bears witness, no less than a success unsurpassed in any later culture in the selection and breeding of plants and cattle. Plato recognized the humane qualities of these simpler communities: so it is significant that he made no attempt to recapture them at a higher level . . . . Plato's

> ideal community begins at the point where the early Golden Age comes to an end: with absolute rulership, totalitarian coercion, the permanent division of labor and constant readiness for war all duly accepted in the name of justice and wisdom.[4]

The word civilization derives from the Latin *civitas*, the root of the word city. So if the first utopia was "the city itself," as Mumford insists, then civilization *is in its essence* a utopian undertaking, an aggressive assertion of organizational, pyramidical power imposed upon and drawing its energy from both nature and the human productiveness of "simpler communities." Once again it is interesting to note that the anarchist tradition in which Roszak places Schumacher goes back, as Roszak says in his Introduction, to "communal, handicraft, tribal, guild, and village life-styles as old as the neolithic cultures."[5] The fundamental root of the eutopian-anarchist tradition, Roszak implies, lies in ancient nonurban social life, and that would place the new ecologically oriented Green politics in the eutopian-anarchist tradition.

By shutting out nature on the one hand (especially stark exposures to death and decay) and on the other by utilizing vast resources of materials and energy to accelerate selected patterns of organizational growth, the civilized human—and most explicitly the civilized man—sought to impose regimes of permanence upon both nature and traditional human culture. The city extracted from nature the raw materials and from "backward" cultures the sheer human energy by which to build, maintain, and expand the empires of civilized permanence. As Senator Calhoun asserted before the American Civil War, slavery of the many enables the civilization of the few. E. F. Schumacher saw quite clearly that chemicalized agribusiness and the deterioration of rural culture were the results of economic processes radiating out of dynamic, industrialized urban areas. In food and other forms of basic production (raw material extraction and handicrafts of all sorts), the industrial revolution was used by capital-intensive economies to supplant peasants, serfs, slaves, indigenous peoples, and small farmers with "energy slave" technologies—agribusiness, above all. What Schumacher apparently didn't understand was that by destroying traditional patterns of cultivation and the complex rural cultures rooted in organic cultivation, civilization was merely expanding its original and archetypal objective: to bring as much of the world as possible (and certainly all of the human world) under the dominion of utopian permanence.

Permanence in the form of Egyptian pyramids or Chinese great walls was, and is, impressive. Because of their massive size, those constructions did give an impression of permanence; they were the materialization of a concept whose construction utilized the most durable of natural objects—stone. But a concept that worked with stone (with what agony and human terror we can only guess) does not work with soil, and one of Schumacher's primary concerns was the deterioration of soil quality under the impact of technological agribusiness. He opened his chapter "The Proper Use of Land," in *Small is Beautiful*, with a lengthy quotation from the work of two highly experienced ecologists—Tom Dale and Vernon Gill Carter—saying past civilizations brought on their own ruin largely by unecological overextension. In their *Topsoil and Civilization*, Dale and Carter insisted "The fundamental cause for the decline of civilization in most areas was deterioration of the natural-resource base on which civilization rested."[6] "Study how a society uses its land," wrote Schumacher, "and you can come to pretty reliable conclusions as to what its future will be."[7]

When the realm of urban *permanence* expands, the realm of natural *stability* contracts, just as synthetic chemicals in soil correlate with humus depletion. The agribusiness chemist seeks only increased yields; the chemical company seeks its yield in profit. Higher yields, in concert with larger and fewer farms, make the farm sector increasingly dependent on market mechanisms while simultaneously providing the vital agricultural commodities by which the permanence of urban-industrial life is encouraged to expand. More and more people come to live in towns and cities, and an urban standard of living becomes normative. The utopian ideal is achieved: effortless affluence in high-rise splendor. Yet we always need to bear in mind that the discovery of horticulture was *the* advance that brought about the emergence first of the settled village, and then quickly, in terms of cultural evolution, the early city. The first farmers were, it is true, domesticating nature, but they did so with a great deal of respect for the inherent *stability* of natural processes. They had neither the machine tool technology nor the advanced consciousness by which to act with unecological arrogance.[I]

Because natural stability—from the Latin *stabilis* or *stare*, to stand, meaning steadfast, steady in purpose, durable, enduring—is both complex

I. For contemporary illustrations of "primitive" agricultural stability, one can turn to the extremely lucid essays on Quechuan farming practices in the Peruvian Andes and those of the Papago and Hopi in Wendell Berry's *The Gift of Good Land.*

and diversified, the minor disruption of a stable, organic whole could only in rare instances cause nature serious injury or threaten a people with starvation. Yet the city's permanence orientation sought to overpower both the fluctuations of nature and the stubborn cultural intransigence of rural rustics in order to create an illusion, at least, of human immortality and omnipotence: the ideal, the perfect, and the permanent. Stability might well lose every battle in the struggle with permanence, just as nonurban cultures were ultimately doomed to lose in real wars with civilization, but permanence can only win the total war against nature by also destroying itself. The devastation of nuclear war is the ultimate apocalypse of utopia, even as globalization correlates to an acceleration of extinctions.

Schumacher was, then, uncharacteristically uncritical in his inclusion of permanence among his "ideals."[II] The establishment or maintenance of permanence is simply outside of, and beyond, human capacity. We can no more create a *permanent* agriculture than we can dig a *permanent* storage pit for radioactive wastes. The time frame is simply too immense; unanticipated variables (earthquakes and ice ages, only two possibilities) are too complex to be forecast. Only sheer utopian arrogance—eco-cultural ignorance posing as civilized expertise—could assert otherwise. What's needed is an ecological dialogue with nature in the renewal of stability. It is the eutopian path that reconnects us with our ancient preurban and preutopian past. To fit ourselves *into* nature may well be a process that is, if anything, even more complex than the civilized rising *above* nature five or six thousand years ago. Agribusiness provides a clear illustration. The present system of food production and distribution relies on chemical, fossil fuel, and industrial inputs to such an extent that if even some of those inputs were withdrawn, the entire system could collapse. (As the Rodale Cornucopia Project points out, transportation may be the most immediately vulnerable factor.) Our present system is the antithesis of stability. Chaos is the polar opposite of permanence; or, more clearly, chaos is the *result* of permanence in a state of disintegration.

II. Schumacher's Roman Catholicism, centered organizationally in urban power and formulating its eschatology in terms of the classical City of God, is itself an expression of archetypal urban utopia. That may be the key reason why Schumacher chose "permanence" over "stability." Stability, as cyclical regeneration, smacks of pagan spirituality. Christianity, with its progressive eschatology leading to the Heavenly City at the end of time, was the principal oppressor of pagan "superstitions" in European—and now global—history.

What we need is a reconstruction (but not a regressive social restoration) of rural stability. This is neither a minor political issue nor an example of semantic nitpicking. Civilization in an industrial form has generated unprecedented alienation from nature and a concomitant atomization of human community. We live, in a sense, within the actualization of Thomas Hobbes' Leviathan: atomized particles held together only by the "terror" of the State. Applied to agriculture, this image depicts raw dirt held together and made productive only by the inputs of urban-industrial technology. Nothing could be further from ecological stability, natural or social. Human communities are held together by natural cohesions of affection and of need, soil in an incredibly rich living matrix of inorganic matter and organic life, plant and animal. Hobbes, the seventeenth-century English political philosopher, was describing the end result of imposed permanence, a projection into political imagery of the underlying fears that the old aristocracy felt in the face of rising democratic demands and expectations. Hobbes' image was one of reaction: a reassertion of the utopian urban pyramid with its elite governing class. Only inputs of terror could keep the destructive wheels of civilized permanence grinding along. What the system defends is not the health, peace, and stability of people or nature; the system defends the perfected *machines* that are its elite vehicles for the perpetual conquest of permanence. According to Mumford:

> The machine that accompanied the rise of the city was directly a product of the new myth; but it long escaped recognition, despite a mass of direct and indirect evidence, because no specimen of it could be found in archeological diggings. The reason that this machine so long evaded detection is that, though extremely complicated, it was composed almost entirely of human parts . . . . [T]he original model has been handed on intact through a historic institution that is still with us: the army.[8]

And "the army" leads us directly to the most pressing disembodied symbol of our present terror: atomic war.

This institution of "disciplined" brutality, whether composed "almost entirely of human parts" or of nuclear and thermonuclear materials ready to envelop the world in atomic ecocide, is a long way from representing the finest or fullest expression of human potential. It is, rather, the product of obsessive greed, entrenched fear, and anxious mortality. It represents a desire for domination and control that is not only unhealthy but capable, by means of modern technology, of destroying Earth as habitat and home. E. F. Schumacher, without the historical specificity of Mumford, nevertheless saw the linkage between environmental abuse and social brutality:

> The social structure of agriculture, which has been produced by—and is generally held to obtain its justification from—large-scale mechanisation and heavy chemicalisation, makes it impossible to keep man in real touch with living nature; in fact, it supports all the most dangerous modern tendencies of violence, alienation, and environmental destruction . . . . In the simple question of how we treat the land, next to people our most precious resource, our entire way of life is involved, and before our policies with regard to the land will really be changed, there will have to be a great deal of philosophical, not to say religious, change.[9]

Although Schumacher erred in his choice of words in the casting of his ideals—selecting "permanence" rather than "stability," he was brilliantly accurate, even prophetic, in his recognition of the dangers in the present system and the potential wholeness of a new order based on proper economic scale and ecological diversity. The unstable and dangerous disintegration of our present economic utopia, with its increasingly belligerent militarism and willingness to jettison civil liberties, unfolds before our very eyes. The eutopia of stability is struggling to clarify its *Weltanschauung* and find its political voice in the global assault that civilization carries on against diverse cultures and nature.

Yet, the emergence of a new stability does not imply that cities should or will disappear. The city *is* a vital organ of human life. (It has brought about the way in which we are now communicating, in all its incredible complexities.) Like Lewis Mumford, we, too, must love and strive to protect the city. But we do so with a heart and mind that love the land and identify ecological integrity as one key measure of cultural coherence. The tasks at hand are to reduce the *size* of cities and to deliberately promote the ecological *resettling* of the countryside. The city compelled a regime of authoritarian compulsion because, in part, it feared country people might well decline the additional burdens of increased production, of compulsive work, of time devoted to surplus labor. But as a civilized people—and we are all civilized now, for better or for worse—we will naturally reconstruct our cities, once given the freedom to do so, once the present utopian system no longer holds us in its paralyzing grip. We now need the city the way we need bread or beer, only we will enter into reconstruction with the city's *cultural* functions first in mind, and with a painfully educated recognition of proper ecological scale.

In the final analysis, cities are also an outgrowth of the Earth. They, too, will stand in a free and wholesome world as fine expressions of beauty, health, and stability. Utopia will have unintentionally brought about an immeasurably refined and magnificently conscious eutopia that reflects, but does not duplicate, the wholeness of preindustrial and preurban life, while simultaneously rising above and transfiguring the social limitations and ecological abuses of traditional patterns of civilized conduct. E. F. Schumacher was, then, an early eutopian whose message for us is both practical and prophetic. He is a wise elder in the youthful deliberations of Green consciousness.

NOTES

1. Roszak, "Introduction," 4.
2. Mumford, *Story*, 267.
3. Mumford, *Interpretations*, 241.
4. Mumford, *Interpretations*, 242–43.
5. Roszak, "Introduction," 4.
6. Dale and Carter, *Topsoil*, 20.
7. Schumacher, *Small*, 102.
8. Mumford, *Interpretations*, 252.
9. Schumacher, *Small*, 114.

2

In Imitation of the Gods

It is my contention that Lewis Mumford's brilliant and concise "Utopia, the City, and the Machine" provides us with exactly that lens through which to understand the linkage between utopia and civilization.

Mumford's essay, included in *Interpretations and Forecasts*, is hugely aided in analysis by *The Myth of the Machine* and *The Pentagon of Power*. What Mumford does is show that the predatory impulse that infuses civilization from its outset—the centralization of political power, class separation, division of labor, mechanization of production, huge military power, slavery, forced labor, exploitation of the weak—has its self-justifying counterpart in utopian fantasy. Not only could "we" build castles, cities, kingdoms and empires out of harsh coercion, "we" were fully justified in doing so, no matter what the human and ecological cost, because "we" were the vanguard of a conception of the ideal, the perfect, and the permanent that might just someday (if we could wrest enough secrets from stubborn nature) enable everyone to live with as much godlike freedom and luxury as "we" do.

The first obstacle to climb over, get around, in this exploration of utopia is the prevailing false use of "utopia," as Mumford clearly underscores in the passages quoted in the preceding essay. Utopia is *not* the alleged mushy-hearted, soft-headed, sentimental, do-good, lovey-dovey, wishy-washy sort of powerless idealism conveyed in conventional usage. Although utopia may have a humanitarian thread running through it, or even a basic humanitarian motive (as in the imaginary world/city depicted in Edward Bellamy's *Looking Backward*), its core thrust is the discovery, creation, and *imposition* of the ideal, the perfect, and the permanent. Utopia is after something in the realm of the gods, of immortality, and its conviction proclaims that this attainment is not only permissible but also possible. The effort is certainly "noble."

The intellectual and even spiritual energy that has thrust humanity and the natural world into the human-contrived artifacts and procedures of technological progress is overwhelmingly utopian: not so much humble tools to ease the labor of our modest earthy subsistence, but brash and brutal technologies unhesitatingly imposed on the natural world to lift us above mere nature, to enable us to live lives reminiscent of aristocracy and even, with a little imagination, in imitation of the gods.

In the February 2003 issue of *The Progressive*,[1] Uruguayan journalist Eduardo Galeano, in a two-page article entitled "Terror in Disguise," asserts that the Market (with a capital M) is a "faceless and all-powerful terrorist that lurks everywhere, like God, and thinks he is eternal, like God." He (the Market) is "an object of fear. He's spent his life stealing food, eliminating jobs, kidnapping countries, and starting wars."

Eduardo Galeano doesn't say so, but the Market is the enforced, extremely high-energy procurement and distribution mechanism of an economy in a compulsory utopian mode.

Karl Polanyi's seminal book *The Great Transformation* is very helpful in understanding the enforced industrial transition, in England, from a basic agrarian, peasant, subsistence economy (with a utopian overlay of aristocratic extraction) to a totally utopianized market economy on an endless, speedy treadmill of abstract production and consumption. Where the king needed peasants, serfs, or slaves to produce basic commodities, the market (with or without a capital M) dissolves folk culture in order to extract the element of labor from an abstract "labor force."

As I argued in *Nature's Unruly Mob: Farming and the Crisis in Rural Culture*, the utopian market corrupts and disintegrates folk culture. We could even say that utopian infrastructure is hardly culture at all but, rather, a weird kind of psycho-technological life-support treadmill system that aims to rise above mere culture, above nature, into the protective encasement of technological immortality. This encasement has brought with it the necessary accrued gleanings of culture—language, symbolic manipulation, accrued scientific insight—but it has, for all practical purposes, both abandoned and extinguished every indigenous cultural root that bears an earthy stain of modest subsistence. We can see this without doubt in the industrial crushing of the peasantry and small-farm class, as well as in the corruption of every indigenous self-sustaining culture worldwide. Civilization, in the Market armoring of its utopian hubris, has gained global supremacy via its undefeatable technological mastery coupled to its truly intimidating arrogance and godlike self-assurance.

This mastery, this hubris, is now heavily congealed in the combined corporate/governmental/military power of the United States of America. This country now seeks—openly so—to be, and to remain, the *only* superpower on Earth. If the Market is the high-energy, intrusive, penetrating mechanism of utopian economic saturation, the military is its "surgical strike" enforcer mechanism, complete with satellites, stealth bombers, rapid deployment forces, "smart" bombs and laser weapons, with an accrued lethality (particularly in the area of nuclear weapons) fully capable of bringing to an end all mammalian life on Earth.

One could say the problem is the military. That is, if we could just find a way either to radically reduce human aggression or, alternatively, find ways to control and contain it that are not ecocidal, then we could all enjoy the benefits and blessings of utopia in peace and security. I believe this view is delusional. The problem of hubris, of utopian arrogance, is by no means confined to the military. The institutionalized military is merely utopia's protector/enforcer arm. With utopia, we are dealing with an ancient human impulse to rise above earthly life, to achieve immortality, to live the life of the gods. This impulse, as Mumford shows (as does Norman O. Brown in his *Life Against Death*), was already in place in the formation of the earliest civilizations.

The issue is not the plucking of utopian overextension from one aspect (the military) of the globalized utopian project; the issue is the radical shrinking of utopia as a whole: the shrinkage of agribusiness, the shrinkage of polluting energy production and consumption, the shrinkage of industrial rationalization, the shrinkage of jet travel, the shrinkage of compulsory schooling, the shrinkage of electronic "entertainment," the shrinkage of intense competitive sports—the shrinkage of utopian presumption and mass acquiescence.

The space vacated by this shrinking needs to be filled, steadily and carefully, with cooperative/communal gardening and farming, renewable energy, durable goods cooperatively produced and utilized, mass transit, sailing ships, buggies and bikes, decentralized ecological education, the rediscovery of folk culture, cooperative play—and the enlarging of earthy modesty and humble engagement.

Utopia is a crime against Creation. We can atone for this crime by living lives of humanitarian servanthood and ecological stewardship. The good garden is eutopian.

NOTES

1. Galeano, "Terror," 18.

3

The Mass-Hallucinatory Fantasy

Barbara Kingsolver, in "A Good Farmer," says she began an earnest study of agriculture sometime around her fortieth birthday. She worked quietly on this project, she says, speaking to almost no one about it, "because of what they might think. Specifically, they might think I was out of my mind." Why would anyone think an interest in agriculture—particularly the small, clean, unpoisoned, homestead sort of agriculture Ms. Kingsolver is engaged in—is a symptom of craziness? "Why? Because," says Kingsolver, "at this moment in history it's considered smart to get out of agriculture."[1]

Allow me—it's a Halloween indulgence—to dress up in an out-of-my-mind costume, bib overalls or peasant's smock, and push open a little farther the battered old door with the creaky-squeaky hinges in the haunted house where the ghosts of agriculture live. In her article, Ms. Kingsolver mentions in passing the Kentucky farmer-poet-novelist-essayist Wendell Berry, whose advocacy of small-scale farming has inspired thousands of people for nearly half a century. Berry is a good and passionate writer. He's written a truly impressive amount of words on the moral, ecological, and political importance of small-scale farming. He has been (and continues to be) an immeasurably great asset to what we might call the agrarian vision, especially when we consider the giant farm stompers, like Monsanto and Cargill, busily crushing all things small in pursuit of globalized "economies of scale."

It is, of course, a truism not only that we live in the present but that we deeply doubt whether history has anything meaningful to teach us. We have supposedly surpassed history. History skulks around in the same demented basement with the ghosts of agriculture. But—hey!—trick or treat; let's at least have a little fun. Allow me to state some flat-footed, boring facts. All biological beings (well, I don't know anything, really, about viruses, for instance, so cut me a little slack about "all")—all biological beings need to eat. We humans are biological beings; therefore—nice, neat syllogism—we humans need to eat. And we eat, mostly, as Barbara Kingsolver points out, stuff that grows in dirt. Dirt is really soil, and soil is largely a blend of minerals with lots of dead and gone, fully decayed, plant and animal remains, all of it the nutrient base for plants whose roots love to feed in this rotten stuff underfoot. Human beings have an ancient relationship with dirt. Without the prior existence of dirt, no mammals whatsoever could have come into being, for without dirt there are no plants, and without plants, no mammals. Pretty simple, really.

So the boring history thing, overlaid onto the boring dirt stuff, may provoke just the tiniest little ghoulish spark of curiosity: are there discernible stages, or plateaus, or revolutionary transitions that characterize the human relationship with dirt? (Remember I have my costume on, this is only trick or treat, not serious, even though, as Ms. Kingsolver says, "If a middle-aged woman studying agriculture seems strange, try this on for bizarre: Most of our populace and all our leaders are participating in a mass-hallucinatory fantasy in which the megatons of waste we dump in our rivers and bays are not poisoning the water, the hydrocarbons we pump into the air are not changing the climate, overfishing is not depleting the oceans, fossil fuels will never run out, wars that kill masses of civilians are an appropriate way to keep our hands on what's left, we are not desperately overdrawn at the environmental bank and, really, the kids are all right."[2]) Just a little spookiness, you know, to stimulate the nervous system. Nothing to be alarmed about, merely the systemic desecration and enforced breakdown of the ecological basis for our existence.

So, are there boring, spooky stages? Of course. Anthropologists probably would point to the use of tools and the controlling of fire as revolutionary transitions. But it's necessary to recognize as well something of a huge consistency in the entire history of human life up through all the gathering and hunting cultures. That is, food was not produced by any sustained process of what we would now call "domestication." Food was gathered. Food was hunted. The transition—the slippery slope we might even say—was provided by gatherers who began to practice horticulture, the deliberate planting and tending of grains and vegetables. This new food abundance increased human population, reduced the need for nomadic mobility, and led—via a combination of depleted local game and enough extra plant food to domesticate the young of select wild critters—to agriculture proper, that is, to an interdependent blend of plant and animal domestication.

Lewis Mumford (*The Transformations of Man*, *The City in History*, *The Myth of the Machine*, etc.) is a wonderful dead-white-man guide into the semi-prehistoric funhouse. (Elise Boulding with her *The Underside of History*, and Gerda Lerner in *The Creation of Patriarchy*, provide guided tours into—*hoo, hoo, hoo*—the *underside* of things.) (Now I hope no one's being frightened here. We're just having fun. "Out of my mind" is just a humorous little turn of phrase. Quit that sniveling right now.) Did I forget to say there *is* a scary part? Sorry, I seem to have mis-shuffled my note cards. Yes, there is a scary part, and we'll get to it in a moment. Please don't scream—it sets my nerves on edge.

It seems that human beings largely thrived on this new abundance, these new cultural revolutions. But something unforeseen happened. In the chapter "Agriculturalization," in *The Time Falling Bodies Take to Light*, William Irwin Thompson says "Just as the collection of grains triggered a runaway positive feedback system that transformed hunting and gathering bands into sedentary farmers, so the establishment of trading and raiding bands triggered a runaway positive feedback system in which matrilineal, gardening societies were transformed into patriarchal, agricultural, militaristic cities."[3] But that doesn't tell us enough; what's more, Thompson zips much too speedily from matrilineal gardening to militaristic agriculture. What happened, of course, is that armed men took control of this horticultural/agricultural abundance, produced kingship, a ruling class, totally subjected the peasant producers, institutionalized militarism and slavery, created "sacred" cities (see Norman O. Brown's *Life Against Death: The Psychoanalytical Meaning of History*, especially—Oh! So wickedly appropriate for our spook house—Chapter XV, "Filthy Lucre"), and generated that historic holy of holies we lovingly call Civilization.

Here's the scary part. Try not to faint, okay? All civilizations have come into being by forcible economic extraction from agriculture above all. Until the industrial revolution, civilization existed explicitly in a two-class universe of aristocrat and peasant, with the former living off the production of the latter. (Serfdom, slavery, peonage—all names for the same basic process.) With industrialization, or almost simultaneously with it, came a number of limited democratic revolutions, beginning with the American and the French. There were, however, major tricks in this treat. The dominant religious thrust in the West was fully fueled by a "salvation" desire to rise above decay, to achieve afterworld immortality, and to happily shed this bag of creepy, earthly worms. And, in a powerfully parallel cultural thrust, the destruction of peasant culture was applauded and encouraged because the ideal for the future, for "progress," was centered on "civilization," and the overwhelmingly dominant imagery for what it meant to live a Real Life derived from slack-jawed awe of the refined, luxurious aristocracy. In the coming world of industrial civility, we would all live as kings and queens, princes and princesses, dukes and duchesses—and no one need ever be a dirty peasant again.

Therefore, commercial agriculture. Therefore, chemicals and artificials. Therefore, agribusiness. Therefore, Monsanto, Cargill, terminator seeds and "free trade" globalization. Therefore, the destruction of the peasantry and small-scale farming worldwide. Therefore, the total penetration of civilization into every possible nook and cranny of the ecological world. But almost nobody stops to think that civilization is, at its core, armed with ecocidal weapons, male-dominated, fiercely determined to achieve and sustain its luxury by means of radically unfair and unequal power arrangements, contemptuous of dirt and those with dirty fingernails.

Guess what? Just like Barbara Kingsolver says, "Most of our populace and all our leaders are participating in a mass-hallucinatory fantasy." This mass-hallucinatory fantasy has a name. It's called Civilization. The thing is, we've all been trained to love this monster, believe it represents our salvation, and cling ever more tightly to its saving principles as things fall apart. So, odd as this may sound, I'm wishing you a truly life-changing fright. Your teddy bear is an ecological rapist who has committed countless crimes against the peasantry. Handle with extreme circumspection. Read Barbara Kingsolver, and in a totally new/old healthy way, go a little crazy. Dress in old clothes. Get some dirt under your fingernails. Get ready for the abolition of this monster. We'll bury your teddy in the garden.

NOTES

1. Kingsolver, "Good," 11.
2. Kingsolver, "Good," 13–14.
3. Thompson, *Time*, 155.

4

The Perfectly Camouflaged Temple

PERHAPS I LOVE MY undereducated analysis too much. Or, to put this little essay, or review, or polemical garbage can on its wobbly trajectory toward a perfectly camouflaged target, I will say while I greatly admire the artful, pithy writing of Arundhati Roy (even regarding communal butchery in India she writes an artful sentence), I am left both stimulated and unsatisfied by a certain slice of analysis in *War Talk*, Ms. Roy's exploration into empire, fundamentalism, and fascism.

One could claim—I suspect critics on the Right do—that Ms. Roy is too clever for her own good, that she would persist being artful, would write her wonderfully clean sentences, even if she dipped her pen in slaughterhouse blood. One could also claim there's something creepily politically correct, over-the-top yuppie, upper-crust intellectual latte, about all this alignment with the poor and dispossessed, all the while being the chic darling of the Starbucks and Barnes & Noble crowd.

Perhaps. We all cast shadows. Selfishness is built into all biological creatures who need to feed their faces and find a mate. But let's not go, at least so soon, off the deep end in regard to Original Sin and human depravity. I rather think of Arundhati Roy as a deceptively delicate-looking, super-strong prehistoric plant, who has thrust her way, sans Miracle Grow, through cement sidewalks, interstate highways, airport runways, nuclear power plant casings and dam footings, to become in an amazingly short time an international medicinal tree, immensely tall, but enormously well-rooted, full of healing fruit. So let's pull up our lawn chairs, in the cool shade of her vast foliage, and chat about a perfectly camouflaged target.

II

We ignore history, and politicians are venal. These are two of the guideposts through which Ms. Roy weaves her literary racehorse. Allow me to quote, in full, two paragraphs—one short, one long—from Ms. Roy's third essay in *War Talk*, "Democracy":

> Historically, fascist movements have been fueled by feelings of national disillusionment. Fascism has come to India after the dreams that fueled the Freedom Struggle have been frittered away like so much loose change.
>
> Independence itself came to us as what Gandhi famously called a "wooden loaf"—a notional freedom tainted by the blood of the thousands who died during Partition. For more than half a century now, the hatred and mutual distrust has been exacerbated, toyed with, and never allowed to heal by politicians, led from the front by Indira Gandhi. Every political party has tilled the marrow of our secular parliamentary democracy, mining it for electoral advantage. Like termites excavating a mound, they've made tunnels and underground passages, undermining the meaning of "secular" until it has become just an empty shell that's about to implode. Their tilling has weakened the foundations of the structure that connects the Constitution, Parliament, and the courts of law—the configuration of checks and balances that forms the backbone of a parliamentary democracy. Under the circumstances, it's futile to go on blaming politicians and demanding from them a morality of which they're incapable. There's something pitiable about a people that constantly bemoans its leaders. If they've let us down, it's only because we've allowed them to. It could be argued that civil society has failed its leaders as much as leaders have failed civil society. We have to accept that there is a dangerous, systemic flaw in our parliamentary democracy that politicians *will* exploit. And that's what results in the kind of conflagration that we have witnessed in Gujarat. There's fire in the ducts. We have to address this issue and come up with a *systemic* solution.[1]

Earlier, Ms. Roy insists that "The Sangh Parivar"—this "hydra-headed, many armed" "'joint family' of Hindu political and cultural organizations"—"understands nothing of what civilization means."[2] Okay, so what *does* civilization mean? I believe Ms. Roy largely wastes her energy looking down an empty tunnel called "secular." Mohandas Gandhi wouldn't have spent a precious minute in pursuit of the worldly to the exclusion of the spiritual. It's a dead end. A totally secular government, totally devoid of the spiritual, is, simply, an illusion. But a *systemic* solution, yes! I submit that the dangerous, systemic flaw in democracy is *the nearly total uncritical attempt to democratize civilization.* In my estimation—I rely heavily on the last several, thick books of the American historian Lewis Mumford (*The Myth of the Machine, The Pentagon of Power, Interpretations and Forecasts*)—the concept of civilization is as fully deified as religious ideology and aristocratic impulse have been able to achieve in several thousand years of serious trying.

This is the core contradiction within our global predicament. On the one hand, civilization connotes by an accrued blend of ideology and mythology the very highest reaches of human attainment. Literacy, literature, critical thinking, astronomy, geometry, mathematics, music, architecture, art, craftsmanship, gourmet cooking, and so on. *Progress* is the modern trademark of civilization: the constant, exhilarating improvement in technology, infrastructure, methods of production, life styles, health care, education, communications, genetic engineering, etc. Until recently, the Left was unambiguously proud to identify itself as "progressive." (One of the best little magazines in America, out of Madison, Wisconsin, still bears the clumsy and misleading name *The Progressive.*) Well, let's take *The Progressive* out of parentheses and simply note that its political agenda is essentially identical to Arundhati Roy's: it is "progressive" in terms of human freedom even as it has slammed on the ecological brakes in regard to economic-cowboy Progress. (Though "cowboy," while cute, is also distractingly misleading. A real cowboy is only a farmer on a horse, and is in no way an apt metaphor for empire, George W. Bush notwithstanding.)

Ms. Roy notes that empire and fundamentalism are linked, above the table, below the table, and when they truly fuse (rather than remain in suspended form, like vinegar and oil), we have direct, in-your-face fascism. The term fascism comes from the Latin *fascis*, that in turn is related to *fasces*. The latter term, in my thick *Webster's*, has these definitions:

> 1. Rom. Antiq. A bundle of rods having among them an ax with the blade projecting, borne before Roman magistrates as a badge of authority. In the fasces of a consul, the ax was included only outside Rome.
>
> 2. The authority or punishment (as flogging or beheading) symbolized by the fasces.

A badge of authority in this context is not the radiant, benign face of a wise elder. It is not a symbol of a democratic council. It is a bundle of rods with ax blade projecting, signifying flogging or beheading. I would assert, leaping provisionally over questions of impulsive human mercy and degrees of authoritarianism, that all aristocracies have been, if meddled with or pushed, fascistic. I would further assert (this is virtually a no-brainer) that civilization from its outset has been an aristocratic enterprise.

Draw the noose tighter on this provisional syllogism and the corpse that dangles from the rope is the peasantry suspended by threads of aristocratic fascism. Or, to be plain and simple: civilization, like the mafia, has from its inception been an exercise in fascist control. If this is true (don't bother refuting me; refute, if you can, Mumford's troubled conclusions), then *the attempt to democratize civilization is an effort, unwittingly, to democratize fascism.* I believe that *this* is the "dangerous, systemic flaw" in democracy that Ms. Roy points at without really identifying it. And I further believe that Mohandas Gandhi, with his unrelenting push for "village uplift" and "village-mindedness," was very close to this realization. The twentieth century—or the twenty-first, for that matter—never faced a bigger threat to the smug presumptions of beatified civility than that snaggle-toothed little brown man in a homespun shawl.

III

Civilization, then, is the perfectly camouflaged target. The core contradiction: civilization as the highest attainment of human ideals and endeavor; civilization always and everywhere a fascistic control system. The industrial revolution opened the castle gates and let lots of people—mostly white Euro-Americans—into the castle precincts to shop the artifacts of neoaristocratic prosperity. Corporations became the new baronage. The machines, fuels, and chemicals of the industrial revolution rendered obsolete overt slavery and crude serfdom. Steps *were* made in the direction of inalienable human rights; steps *were* taken toward democratic self-governance. Perhaps there even was a Golden Age in the New England-style village-square democracy in the nineteenth century—provided you were white and male.

But the octopus of triumphant Empire forcibly infiltrated every nook and cranny of decentralist folk culture and mutated allegiance onto itself. The not-yet-slaves, via commercial allurement and patriotic propaganda, are tugged into Empire obedience by school, workplace, entertainment site, sporting event, shopping mall, and place of worship. In this process, civilization is the perfectly camouflaged temple of fascism—fully in the open, brazenly self-aggrandizing, its bloody origins gilded with mythology. Civilization is *always* good, *never* evil. It represents our successful elevation above primitivity, coarseness, discomfort, backwardness, hard work, dirt, and—by linking with an elemental human desire for the perpetual maintenance of self—mortality. In this context, with this huge accrued load of ideology and mythology supporting and maintaining the deification of civilization as our savior from death and the dreadful past, our task, in the face of the magnitude of this present Empire threat of ecocide and mammalian annihilation, is even larger than Arundhati Roy describes.

We not only have to resist Empire and, possibly, defeat it—whatever that may mean; we also have to abandon—deconstruct—the perfectly camouflaged temple of civility. Without this purging, the dangerous, systemic flaw will continue to plague us—plague us, that is, until either Earth is effectively destroyed or an explicit and very hard-edged aristocracy restores us peons to a hereditary condition of peasant servitude, gloating that, iffily nip and tuck for a century or two, even explicit democracy could not budge our conditioned need to barter freedom for a Leader or eutopian sufficiency for utopian surfeit.

IV

Let's go back to Arundhati Roy. In the two paragraphs quoted earlier, she said the "dreams that fueled the Freedom Struggle" had been "frittered away like so much loose change." The Freedom Struggle was, of course, the Gandhian liberation movement. Ms. Roy goes on to say politicians and political parties have "tilled the marrow" of secular democracy, mining it for electoral advantage, made tunnels and passages, undermining the meaning of "secular." All this has "weakened the foundations of the structure," the "configuration of checks and balances." There's fire in the ducts. I don't dispute the truth of this—we have seen a similar process in the George W. Bush administration—but I object to its incompleteness, its half-truthness.

The Indian Freedom Struggle cannot be separated from the person of Mohandas Gandhi. Mohandas Gandhi cannot be separated from the principle of nonviolence, and Gandhi's nonviolence cannot be separated from his intense and all-pervading spirituality. Ms. Roy starts with frittered-away Freedom Struggle dreams but ends up bemoaning the undermining of "secular." Something got lost, dropped like a used kleenex, in this process. What is it? In broad terms, the answer is easy. The *spiritual vision* that was at the center of those dreams got lost, dropped, and frittered away. To ask what the spiritual vision is, what it really consists of, how do we live it out—that's what's hard. Gandhi's genius (if we might call it that) lay not in separating the secular and the spiritual, but in fusing them. If anyone misses this point, so elemental in any analysis that involves Gandhi, the subsequent conclusions are bound to be misleading, even dead-end.

Two Gandhi quotes—they could be piled on endlessly—as taken, almost at random, from Thomas Merton's *Gandhi on Non-Violence* illustrate this point:

> On India rests the burden of pointing the way to all the exploited races. She won't be able to bear that burden if non-violence does not permeate us more than today . . . . India will become a torch bearer to the oppressed only if she can vindicate the principle of non-violence in her own case, not jettison it as soon as independence of foreign control is achieved.
>
> Jesus was the most active resister known perhaps to history. This was non-violence par excellence.[3]

Merton himself, in his long introduction ("Gandhi and the One-Eyed Giant") to this acute selection of Gandhi's pithy statements, emphasizes the "inherent relation between non-violence and the renewal of India for which Gandhi lived and died." Here are additional passages from Merton's introduction that underscore the point:

> The highest form of spiritual freedom is, as Gandhi believed, to be sought in the strength of heart which is capable of liberating the oppressed and the oppressor together.
>
> The only way truly to "overcome" an enemy is to help him become other than an enemy. This is the kind of wisdom we find in Gandhi. It is the wisdom of the Gospels.[4]

V

The temporary advantage of sticking with "secular," to the almost total exclusion of "spiritual," is the latter so often (and so quickly!) mires us in exactly those so-called religious passions that are the bane—the Sangh Parivar, the Moral Majority—of thoughtful, considerate, humane governance. Religion is therefore to be avoided as totally as possible, a bubbling cesspool of superstitions, irrational convictions, and dangerous impulses that are bound to (and intend to) undermine the secular and concentrate power in theocratic or semi-theocratic top-down governing structures. So, to defend the secular, we eschew the religious. But Gandhi, the icon of our vision, was an intensely spiritual man. Fasting, praying, hymn singing, and frequent allusion to God were all part of Gandhi's life. Go ahead, make sense out of this very serious mess, untie this Gordian (or Gandhian) knot without using Alexander's sword, resolve this conflict by means of nonviolence.

The first clue is a hard one. If Gandhi achieved the great Freedom Struggle by means of fusing the secular and the spiritual, so we are going to have to unravel that seemingly impenetrable knot of secular/religious contradiction in order to rebraid it into something humanely and ecologically useful: Green politics in a Green culture. What's the knot? I submit the knot is civilization itself, and that religion is, pretty much as secularists recognize, the hidden threads, the submerged fibers, that keep the knot locked. And, as far as I can see, the key—or a key—element in gluing the Gordian knot in a seemingly unbreakable braid is the dominant conception, understanding, and image of God. God is the glue in the Gordian knot.

Without recapitulating all the background history presented in the preceding essays, let's merely note that what Marcus Borg, in *The God We Never Knew*, calls the "monarchical model" of God was, "roughly from the fourth century," the model that "has dominated" in Christian teaching and preaching:

> Because an earthly king is male, God is imaged as male. The image of God as king is not only anthropomorphic but andromorphic (andros is the Greek word for male) . . . . Like an earthly king, God is powerful, but he is even more than that: God is all-powerful—that is, omnipotent. Like an earthly king, God is the lawgiver and judge, but with even more serious consequences. Our well-being in this world—and, in later layers of the biblical tradition, our eternal destiny as well—depend on our observance of his law.

> Finally, the image of God as king suggests that God is distant. The distance between an earthly king—politically, economically, and architecturally, as well as in power, wealth, and lifestyle—and peasants, the ordinary people of the time, was immense.
>
> In relationship to this image of God, who are we? What image of ourselves goes with God as king? We are subjects and therefore "not much." In relation to God as king, we are peasants....
>
> The monarchical model shapes not only the internal dynamics of the Christian life but also Christian perceptions of nature, politics, and gender. Its central image of domination and subjection goes with the domination of nature by humans, the domination of other human beings by political and economic elites, and the domination of women by men. It is not the singular cause of any of these, but it is part of the ideological package that has legitimated domination....
>
> A monarchical model of God and a monarchical political order go together. To state the obvious, the metaphor of God as king originated in such societies. But more is meant. Namely, the monarchical model of God has most often functioned to legitimate what Walter Wink calls "the domination system," the most common form of political society over the course of recorded history. Domination systems are hierarchical social orders marked by economic exploitation and political oppression in which a few people (almost always men) rule over everybody else.
>
> Though the image of God as king can be used to challenge domination systems (as noted earlier), more commonly it legitimates the earthly king's rule and the social order over which he presides. God as king is the source of the society's values, laws, and structure, and the king rules as the vice-regent of God. Microcosm mirrors macrocosm; the earthly king's rule replicates the rule of the divine sovereign. The king is the number one figure in the domination system, and his role at its pinnacle reflects the will of God. Chosen by God or anointed by God, the king rules by divine decree, a status frequently ritualized by the participation of religious officials in the king's coronation. The result is a domination system legitimated by God.[5]

The passages just quoted are all from Chapter 3, "Imaging God: Why and How It Matters." Borg waits until Chapter 6, "The Dream of God: A Politics of Compassion," to tell us more about the internal workings of such societies where the monarchical model of God has most often functioned, and about how the "Bible is a tale of two kingdoms: the kingdom of God and the kingdom of this world."[6] Borg may be a bit pious, however, with his "tale of two kingdoms." John Dominic Crossan, in his *God and Empire*, says depictions of God as violent and God as nonviolent "run side by side, and often in the same books, from one end of the biblical tradition to the other."[7] If Crossan is right—and it's obvious he is—it's not just kingdom of God versus kingdom of this world, but violent God versus nonviolent God. Let's indulge in a rather long quotation from Borg's book, even as we recall that we are still looking for Arundhati Roy's loose change:

> In order to see the tension between these two kingdoms, these two ways of being and living in community, we begin with a picture of the most common type of society (including social, economic, and political dimensions) in the world in which ancient Israel and the Bible came into existence.
>
> This was the social world of the "preindustrial agrarian society," sometimes abbreviated as "peasant society." This type of society is easiest to envision when we place it within the sequence of human social development. Preindustrial agrarian societies began about six thousand years ago. They were preceded by simple horticultural societies in which agricultural production (using digging sticks and hoes) was modest, and only small concentrations of population (at the most, villages and towns) could be supported in one place. Preindustrial agrarian societies emerged when agriculture became sufficiently productive (through the use of the plow) to generate surpluses that could support larger concentrations of population. In short, cities were born.
>
> Preindustrial cities were small by modern standards. On average, 10 percent of the population lived in cities, while the other 90 percent remained rural. The rural population was made up primarily of agricultural producers (peasants, from which we get "peasant society"). In such societies, agriculture was the primary source of wealth. There was as yet no industry; manufacturing was by hand and small scale. This type of society lasted in various forms until the beginning of the industrial revolution, and it persists in some developing parts of the world to this day.
>
> Of greatest importance for our purpose is the social structure of preindustrial agrarian societies. The division between city and country corresponded with a class division of power and wealth between urban elites and rural peasants. The urban elites consisted of two groups. First there were the elites proper, typically 1 to 2

percent of the total population: the ruler, aristocrats, high government and religious officials, and their extended families. Second, attached to the elites was a service class known as retainers, typically 5 to 8 percent of the population: lower-ranking government officials, soldiers, priests, scribes, urban merchants, and so forth.

The economic gulf between urban elites and rural peasants was enormous. Elites and their retainers (together, less than 10 percent of the population) acquired about two-thirds of the society's annual production of wealth, with about half of the total going to the top 1 or 2 percent. Rural peasants (90 percent of the population) made do with the remaining one-third.

How did the urban elites mange to do this? They didn't produce anything. They provided few services (going to war being one of their chief ones). Rather, they got their wealth from the only source of wealth there was: the agricultural production of peasants. The elites used two primary means of extracting wealth: taxation of peasant production and ownership of agricultural land (which could then be rented by peasants, or farmed through the use of laborers or slaves). Elites were very good at calculating exactly how much they could take without driving peasants into starvation or resistance. This is not surprising, for their wealth depended on this relationship of economic exploitation. Moreover, wars were often fought between elites in order to acquire control over more peasants; this was the only way in which their wealth could be increased. The cumulative effect on peasants was dreadful. Peasant existence was vulnerable to subsistence diet, drought, illness, war, even the death of an animal. Life expectancy was very low.

In these societies, religion most often functioned to legitimate the social order: God had ordained that it be this way. At least the religion of the elites did, as it has come down to us through their retainers of priests and scribes. For the most part, we do not have access to the religion of peasants, who left few or no written records.

Thus, "peasant society" was the dominant type of social organization in the world in which the Bible came into existence. City-states and eventually agrarian empires emerged in Egypt and Mesopotamia beginning around 4000 B.C.E. By the time of ancient Israel's origin, this type of society was firmly entrenched in the ancient Middle East. Ruled by elites, such societies were intrinsically hierarchical, economically exploitative, and politically oppressive.[8]

What is really most striking about this passage—even as the data hammer home the magnitude of exploitation by "urban elites"—is that the overall society is called "peasant," and that the two words that would far more accurately describe or identify the exploitative structure of this society are never uttered. Those words are "aristocratic" and "civilized." Borg tells us "The dream of God—a politics of compassion and justice, the kingdom of God, a domination-free order—is social, communal, egalitarian. But our dreams—the dreams we get from our culture—are individualistic: living well, looking good, standing out."[9] And, while "individualist" has a thread of truth in it, the deeper reality is conveyed by Lewis Mumford in Chapter 10, "The Burden of 'Civilization,'" in *The Myth of the Machine*:

> The underlying assumption of this system is that wealth, leisure, comfort, health, and a long life belonged by right only to the dominant minority; while hard work and constant deprivation and denial, a "slaves' diet" and an early death, became the lot of the mass of men.
>
> Once this division was established, is it any wonder that the dreams of the working classes throughout history, at least in those relatively happy periods when they dared to tell each other fairy stories, was a desire for idle days and for a surfeit of material goods? These desires were kept from an explosive eruption, perhaps, by the institution of occasional feasts and carnivals. But the dreams of an existence which counterfeited closely that of the ruling classes, as the brummagem jewelry worn by the poor in Victorian England imitated in brass the gold baubles of the upper classes, have remained alive from age to age: indeed they are still an active ingredient in the fantasy of effortless affluence that currently hovers like a pink smog over Megalopis.[10]

When Marcus Borg asked how the urban elites and their retainers managed to take "two-thirds of society's annual production," he says it was by taxation of the peasantry and by ownership of land. Well and good. But why did the peasantry obey and comply? Here we are back to the knot. The peasants complied because they were coerced, by force of the sword, and they were overawed by the triumphant, self-assured broadcasting of a monarchical God whose will for the common people was obedience to the rule of a "divine" (or divinely installed) sovereign. They were intimidated and browbeaten by the monarchical model of God; and, in the long,

long process of this conditioning in enforced deprivation—hundreds and even thousands of years in the making—when the opportunity arose (the industrialization of production), the lower classes were both forced and lured out of their relative cooperative culture of subsistence into a system of mass consumerism (Mumford's "pink smog") that required a shift in obedience from peasant servitude to factory wage-slavery. The elite managers of "democratic" consent have striven mightily—and, thus far, successfully—to keep our collective understanding of democracy, of what a "democratic society" might really mean, at this level of illusory suspension.

We are obstructed from even clearly *wanting* a "domination-free order" because of our deeply ingrained patterns of deference to the king, to the aristocracy, to urban elites and to the monarchical model of God, that keeps this package of psychocultural deference sufficiently glued to our desires, our dreams, and our consumerist fairy stories. The god whose boiled-down glue keeps the Gordian knot intact is the fictitious god of kingship, of divine right and Manifest Destiny, without which civilization, as a domination system, unravels. And we are all psychoculturally addicted—*spiritually* addicted—to the "security" of the domination system.

In addition, a great many people who think of themselves as totally "secular" have an acculturated bias against the past, against "backwardness," against agriculture and the peasantry. Real life is urban. The rural past is a kind of historic prison record that is to be forgotten as quickly and erased as fully as possible, and not at all to be seen as a tormented and abused model that—if we are to have a Green future worthy of the name—requires both resuscitation and resurrection. Without this grounding in nature, without this commitment to ecological living, there is insufficient traction, intellectually and spiritually, to deconstruct the domination system. Green politics requires Green culture, and Green culture is created by breaking with the domination system, by "voting" with our lives. Environmental technofix is largely a dabbling avoidance. This avoidance enables "secularists" to continually embrace "progress," and the chief unifying concept for progress is "civilization." The aristocratic Right loves civilization for obvious reasons (civilization is the creation and political plaything of aristocracy); but a significant slice of the Left, otherwise and deliberately devoid of a unifying god concept, has for all practical purposes deified civilization itself as the primary symbol of desired attainment. This secular deification of civilization is the voluntary nose ring by which the Left consents to be jerked around by the Right.

Civilization always and everywhere has been a "domination system." No civilization anywhere has ever been anything but a domination system, although the magnitude of extraction and intensity of oppression have varied greatly, and industrialization has, to some extent, shifted the burden from sheer human labor to an increasingly threatened global ecology. We would recognize immediately the absurdity of attempting to "democratize" slavery or the aristocracy, but we can blather on endlessly about democratizing civilization, even though civilization in its monarchical modeling, in its inherent elitist structure, is a system of entrenched exploitation. Civilization, at its core, has never been anything but extractive and oppressive.

We will be ready for what Marcus Borg calls the "dream of God" when we are willing to abandon our allegiance to the perfectly camouflaged temple we call civilization. There is a god hiding in civilization's glue, and we will wake from our sniffling intoxication, our toxic intoxication, only when we are truly liberated from this evil god. What Marcus Borg calls the "dream of God" is what Lewis Mumford calls "eutopia." Green politics is the eutopian dream of God.

NOTES

1. Roy, *War*, 39–40.
2. Roy, *War*, 25, 36.
3. Merton, *Gandhi*, 40.
4. Merton, *Gandhi*, 14, 15.
5. Borg, *God*, 63, 68, 69.
6. Borg, *God*, 134.
7. Crossan, *God*, 83.
8. Borg, *God*, 134–36.
9. Borg, *God*, 145.
10. Mumford, *Myth*, 214–15.

5

Preliminary Thoughts on Green Education

If the purpose of Green politics is to encourage and help create the conditions for ecological living on Earth, how would we describe the purpose, function, and daily routine of Green education? What would an elementary school, for instance, look like in a truly Green culture? I would like to wander, a little, in Green imagination in regard to schooling, but first I want to sketch the intellectual framework that gives shape and guidance to this wandering.

Once of the thinkers I most rely on is the late Lewis Mumford, probably the most prophetic of American historians. Mumford was an American, but he was a world historian whose grasp of historical dynamics was so deep that he was able to project—for instance in *The Pentagon of Power* (1970)—the incremental tightening of the "megamachine" noose in the late twentieth century. In other words, he anticipated our current disaster. Let me take a moment to sketch Mumford's intellectual framework without, I hope, doing too great an injustice to its complexity.

In "Utopia, the City, and the Machine," an essay reprinted in *Interpretations and Forecasts: 1922–1972*, Mumford says that a compulsion for utopian control, utopian perfection, is inherent in the earliest civilizations: "pyramid" regimes typified by armed and deadly elite control, institutional warfare, systemic slavery, the oppression and expropriation of agriculture, a drive to "conquer" nature, and the religious celebration of probing and controlling male intelligence. In *The Pentagon of Power*, Mumford shows exhaustively how political conquest and scientific ideology from roughly the fifteenth century onward coalesced in the "megamachines" of the twentieth century, and how pervasively totalitarian and antinature these political megamachines really are.

The military, of course, is an ancient aspect of this totalitarian, pyramidical system. Two other aspects now infest our lives, our culture, and our consciousness, but it takes a historian with the depth of understanding of a Mumford to clearly identify them and to recognize their deadly megamachine features. These two aspects are the factory and the school; that is, the sort of social regimentation that *never occurred* in human culture until early civilization invented and imposed it—slavery and the military, in its most institutionalized formations—is now the norm of our everyday lives. Utopian civilization has simply saturated the world with its compulsory regimentation. Slavery is both mitigated and disguised by the wage system (therefore compulsion in the workplace has a nominally voluntary feature), but property and money make noncivilized subsistence virtually impossible, as Indian reservations demonstrate.

As with all complex patterns of thought and analysis, Green thinking has many gradations. "Helping the student learn about nature" has a Greenish tinge to it, and my limited exposure to contemporary curricula suggests there's a fair amount of Green-around-the-gills instruction going on: global warming, glacier melting, species extinction—that sort of thing.

My purpose here is not to critique the scientific veracity of this teaching but to look, both explicitly and implicitly, at how the standard school system *embodies* the overbearing compulsory features of the megamachine and is the institutional vehicle in which and by which our children are trained, conditioned, and programmed to fit in with the utopian economic and military megamachine later in life. The school, despite elements of intellectual content to the contrary, is where we all were, and are, conditioned for utopia, just as the factory is where the near-perfect artifacts of utopia are fabricated, by alienated labor, for the utopian standard of living whose energy requirements have now generated a massive climate change over the very Earth on which we've evolved over hundreds of thousands of years. This is a feedback loop with an attitude.

I was raised on a small, first-generation homestead farm in north-central Wisconsin, a mile or two west of that river that has carried our state's name and, thanks to paper mills, a sizeable quantity of our processed forests' wastes. The garden, farm, and woods provided most of our food, all of our firewood. Our neighbors, the forest, and a nearby little river provided the bulk of our "entertainment." I went to both one-room schools in the township—one room for all eight grades—and do not wish to idealize these schools, but there were features of one-room schools—the small size, the neighborhood proximity, the community function they served—that were good, desirable, and worth preserving. The teaching and curricula were, let's say, early utopian: already pointing in the direction of white-collar success, abstract knowledge, mindless routine and psychotic compulsion, almost totally indifferent toward farm life, rural culture, and the natural world right outside the windows.

With school consolidation in the early 1960s, and the coming of the computer messiah in the 1990s, these utopian characteristics have greatly deepened and become more totally normative, while the latent Green tendencies of the one-room schools, never adequately recognized and not at all protected, have almost disappeared from cultural consciousness. A school without yellow busses, a swimming pool, a principal, and an absolutely predictable routine is virtually unimaginable—a fantasy totally ungrounded in utopian "reality."

So what would a Green school look like? The local district has a school forest and a school forest lodge. Lots of kids use the facility. Some classes get to stay overnight. And while I do not wish to disparage any school forest program or activity, I have yet to meet a student in or graduate of the local system who can go to the woods and easily identify tree types, wild flowers, bushes, bugs, or habitat—or, for that matter, hardly anyone who simply loves to wander in the forest.

I suspect the primary response to this observation is, largely, "So what?" This is a most marvelous place to begin a little intellectual spelunking, a little wiggling into and through the unacknowledged chasm between our loud and boastful emphasis on Science (with a capital S) and our almost total indifference to the actuality of the natural world that surrounds us. Science, of course, means Chemistry, Physics, Mathematics, Computers. It means sustained immersion in an approach to, and attitude toward, the natural world that is based on reductionism and manipulation. With the formaldehyde-smelling ambiguity of Biology to the side, our entire approach to the natural world is overbearingly abstract and geared toward a practice of Science funneled into technology and the ever-deeper penetration of nature's inner structure—not for tender and reverent knowledge, but for cool and greedy power.

Mumford's *Pentagon of Power* shows how the drive for human domination of nature, at least in its modern form, derives heavily from two trends easily traced to Europe from the late fifteenth century onward. Those trends are outward territorial expansion—world conquest and colonialism—and an increasingly congealed scientific-technological-industrial drive to reshape nature and the human community in civilization's own abstract, utopian image. Here's Mumford from page 4 of his *Pentagon*:

> One mode of exploration was concerned with abstract symbols, rational systems, universal laws, repeatable and predictable events, objective mathematical measurements: it sought to understand, utilize, and control the forces that derive ultimately from the cosmos and the solar system. The other mode dwelt on the concrete and the organic, the adventurous, the tangible: to sail uncharted oceans, to conquer new lands, to subdue and overawe strange peoples, to discover new foods and medicines, perhaps to find the fountain of youth, or if not to seize by shameless force of arms the wealth of the Indies. In both modes of exploration, there was from the beginning a touch of defiant pride and demonic frenzy.[1]

These touches of defiant pride and demonic frenzy now assert themselves as the only viable path of rationality—witness the current either/or of civilization versus terrorism—that casts all who would question or challenge, much less defy, civilization's absolute and inherent goodness as either foolish or evil. We drift along with conceptions of utopian good versus natural evil that direct our lives in innumerable ways. Even those of us alarmed by forecasts of global warming disasters still hop in our cars to go do trivial things. We watch and listen to reports of two million-plus inmates in American "correctional" institutions, including Wisconsin's very own Supermax. We read about the ever-increasing concentration of corporate power via merger and acquisition. We know that scores of billions of dollars are spent each year on military force to "protect" "our" oil supply from the Middle East. We have some kind of handle on cognition in regard to deindustrialization, the nearly total destruction of small-scale farming, drug consumption (and the "war on drugs"), the deepening divide between the haves and the have-nots, sweatshop labor and brand names, and the extermination of indigenous cultures worldwide.

Every slick and glossy ad on commercial television and in high-end national magazines tells us to celebrate power, indulge in perfection, consume without limit or consequence, believe that technological intelligence is in control. Every school system chugs along, doing its own little optimistic part—thanks to property tax and an ersatz "property tax relief"—to "meet the needs of the twenty-first century." Just as the truant kid cannot possibly have a rational and justifiable reason for seeking to evade the compulsory utopia, so too "terrorists" cannot possibly have legitimate grievance against "civilization." That a truant or troubled kid can become a "terrorist"—witness Columbine—only serves to activate our conventional fear, multiply security measures, intensify our need for snitches, and forge closer links with less and less civil liberties scrutiny between the compulsory utopia of the school and the compulsory dystopia of the jail. Witness John Ashcroft and Alberto Gonzales.

At what point do we read not only Mumford but also Paul Goodman (*Compulsory Miseducation*) and Ivan Illich (*Deschooling Society*)? No, we don't read them, we can't read them—or, if we do, the analysis we find there is so "outside the box," so demanding of a social, economic, political, cultural, and educational reformation that it all seems absurd or hopeless. So we drift, incremental step by incremental step, toward global disaster. We *know* this is true. But. But is such a wonderful, all-purpose word. It's a post sunk to the center of the world around which we can change direction at whim or spin, spin, spin. Isn't it simply glorious that we have an administration that is in process of ridding the world of evil? Such fine frat boys, such unparalleled country club rug rats, such pious little squeaky-clean Sunday school snots—all of them tinkering with the controls to apocalypse. God's little cheerleaders with fifty-star pom-poms.

So—what about a Green school? To think about a Green school and a Green culture requires either hope—a conviction that somehow we will more or less safely emerge from this intensifying global hostility and carnage—or it requires escapist fantasy. There are people who, in principle, are pushing in a Green direction. There's a new private school called Conserve near Land O'Lakes, Wisconsin, and a public charter environmental school in Wisconsin's Kickapoo Valley. I don't frankly know enough about either to say much, except vaguely point in their direction. And yet, a Green school—probably either private or public charter—*is* possible, provided a group of capable, persuasive, and politically relentless people really wanted one in their community. The evidence and the contradiction in regard to science—our absolute and uncritical wallowing in abstract science and our indifference toward and contempt for non-invasive natural science—is palpable. A determined community group could embarrass a moderately enlightened school board into a creative act, provided *the money* could be found—given the superintendent's salary, the building debt, the retirement fund, and all that. The football team. The budget crunch.

A small school, built largely of local and native materials (logs, lumber, field stone), carefully sited to both fit the landscape and be close enough for at least a fair proportion of the kids to walk or bike to—this is a starting point. In the woods, near a river, creek, or lake. Garden space and compost bins. A kitchen and a workshop. What's to prevent a school being multiple-use? Simultaneously, a new town hall, local library, community meeting facility, theater, even office space for county, state, federal, or international agencies? An office of the United Nations?

Get the kids *out of the classroom*. Outside. Boots and raincoats. Snowshoes and mittens. Wading shoes and small backpacks for lunch, flora and fauna identification books, simple equipment (like a magnifying glass). Theater and crafts and languages and field trips to—Oh, my God!—Indian reservations and—double Oh, my God!—student exchange programs with Indian nations and inner-city kids from—triple Oh, my God!—Milwaukee. Does a kid want to stay at home today to help her dad bake bread? Encourage such behavior rather than punish it. Does somebody have an intense crush on Elizabeth Cady Stanton or John Muir and just can't quit reading? Let them alone except to pile on more books. (But ask for a class report, an essay, a lecture, a drawing.) Is there someone who wants to snowshoe to a remote spot on the river and just sit all day, in a prebuilt blind, with notebook to record what happens inside and out? Show that person the journals of Annie Dillard, Thomas Merton, and Henry David Thoreau. Toss in some haiku poetry. Encourage such concentrated inwardness.

Is it really absurd to believe that a sixteen year-old can weave a rug, build a shed, can vegetables, tend a garden, identify at a glance virtually every tree in the woods, recognize cloud types and weather patterns, feel comfortable in the forest, play one or more musical instruments, help out at a nursing home, speak at least one "foreign" language, clean a chimney (or build one), split firewood, sing folk songs without false posture or emotional paralysis, milk a cow, know elemental astronomy and geology, make cheese, and canoe the full length of the Wisconsin River in order to graduate? Well, throw out whatever of that doesn't appeal to you and make your own list—or would you rather have your children and grandchildren and great-grandchildren just keep "going to school" so they can "graduate from college" so they can "get a good job" and "make lots of money" so they can have a "high standard of living"? Why not just buy them caskets and cemetery plots and not waste so much effort?

Are we alive? Are we citizens? Or are we conditioned two- and four-year suckup dependable voters, choosing the always-increasing lesser evil, never daring to exceed our precious little comfort zones? Our complicity in this lethal utopian system is heartbreaking as we mouth all the poisonously pious platitudes about "freedom," "free trade," "free enterprise," "free will," "God's will." Or "security." There are still two ethical blocks to stand on, maybe only two. One is stewardship in Creation, and the other is servanthood among our own kind. If these ethical blocks have colors, the first is Green and the second is Red—stewardship and sharing. If we are to have a future worth living, these are the colors we will live by. Otherwise, as in C. S. Lewis's *The Great Divorce*, it's the grey city and a long, long wait at the purgatorial, utopian bus stop for the bus that never arrives.

This bitter little essay is not what I set out to write. My intent was to describe, as fully as my blighted and meager imagination could achieve, the simple and complex beauty, the peacefulness, the neighborliness, the sense of sufficiency, the attentive sensitivity toward the natural world that surrounds us: how utterly *wholesome* it would be to live in a cooperative, self-provisioning gardening-and-farming community with cottage industry, elder hostels, homey recreation lodges, inns, rustic assisted-living facilities, a Buddhist/Quaker monastery—and a little school, down by the river, in which the keeping of routine, punctuality, and test records was low on the list of educational priorities—something at once very folksy and sophisticated, simultaneously local in an enveloping way and yet globally attentive in a caring, compassionate way.

I begin to imagine I understand the sorrow, dismay, and incredulity of Jesus when he told people the kingdom of heaven *is at hand*: let go of your pride, your greed, your neurotic security obsessions, your privatized salvation preoccupation, and engage the life that lies richly and cooperatively at hand. Like Christianity that has made a total otherworldly salvation cult out of Jesus' "kingdom of heaven" (emptied almost completely of earthly content), so I fear that "Green" will by hyped, when it is inevitably grabbed by the political parade types, as economistic technofix—"natural" capitalism. Gag me with a solar panel. And "Green" education will be how to achieve the Los Angelization of the planet with a recycled minimum of "externalities."

Meanwhile, some of us *do* have a different vision with glimpses of simple, grounded earthiness from Martin Buber's "community all through," Wendell Berry's longing for authentic settledness in undefiled landscape—even some literary images from C. S. Lewis's Narnia and J. R. R. Tolkien's Hobbit Shire. And for those who snort and sneer and shake their heads at this alleged pseudo-innocence, this supposed inability to recognize the pervasiveness of *evil*, we reply, with sorrow: we are looking at evil, we were raised in evil, we were solicited and drafted by evil, we have been trained in the rationale and dissemination of evil, we are "protected" by evil. This evil's name is civilization. God save us from this utopian evil.

NOTES

1. Mumford, *Pentagon*, 4.

6

Suspended in Civilized Values

We are—perhaps not fully—caught in a powerfully complex cultural suspension, although it's hard to generalize about how much or how fully people glimpse, much less grasp, the extent and magnitude of this suspension. As children, we experience the world as the world is. The way things are is a given. It appears that, for most people, this sense of givenness persists throughout life. The way things are is the way things are; and, although we may feel dissatisfaction, uneasiness or anxiety, to imagine (much less believe in) a radically different future reality is either amusing or flaky. Or both.

We may admire, perhaps even secretly envy, the people who raise most of their own food, who refuse to have a television, who keep their kids out of school, who experience childbirth at home, who live with an aging, ill or handicapped person, who heat and cook with wood for fuel, who generate their own electricity by wind or solar (or who decline electricity altogether), who practice meditation or espouse a life of home-made simplicity, who sew, do carpentry or stone masonry—who decline, in short, to be as fully hooked into the system of prevailing institutions as most of us, and who demonstrate their unconventional disinclination by engaging in determined self-reliance. Broadly speaking, this unconventional self-reliance is the scattered, tattered remnant, the cultural residue, of a deep and broad "self-reliance" practiced almost universally only a few generations ago. What until recently was the norm is now a stubborn cultural oddity, an atavism.

Perhaps it's true that all culture, certainly including the language that glues us to our culture, is inherently restrictive. We are shaped and funneled in certain ways distinctive to the circumstances of our cultural upbringing. It appears that this shaping, though unique in its particulars, is a cultural universal. In this country, we are expected to go to school, get a job, get married, raise a family, live a "normal" middle-class life. Most of us do. We vote A or B, Republican or Democrat, watch the network news, maybe go to church, hope that the economy stays "strong" and "healthy," deplore the mindless hatred of those evil terrorists who hate our freedoms and superior way of life.

Where are we suspended? We are suspended in a huge fabrication of "health," "freedom," and "salvation." The economy is, largely, a mindless parasite, a toxic parasite, on Earth's ecology. What is "health" for us is destitution for millions of radically underpaid and chronically oppressed sweatshop and agricultural workers the world over. Both our economic and political systems are enforced by the world's most massive and deadly military in history, and our bland contentment and resentful sense of entitlement dovetail exquisitely with an inherited doctrine of exclusive religious superiority. We are free, healthy, and saved, while the bulk of the world exists in unfreedom, unhealthiness, and in likely damnation because (the reasons blend into one another) the people there are caught in backwardness, are lazy, are bigoted, are not very bright, are entrapped by false religions, have skin colors and cultural liabilities that virtually guarantee their perpetual foolishness and inferiority. That is, our very sense of moral superiority and religious entitlement—all justified by an obvious set of material advantages—locks us into a cultural suspension. To become, to whatever degree, unsuspended is therefore risky business. It is politically, religiously, educationally, economically, perhaps even medically against the grain. It is rebellious. It is to leave the obvious security of the cultural herd for the uncomfortable loneliness of the iconoclast.

II

All this is why the Democratic Party is locked into being Republican Lite. Let's just start with Earth's ecology, global warming in particular. It's a simple truism that a huge reduction in fossil fuel consumption—oil and coal especially—is a key and critical step in addressing global warming and climate change. But first it needs to be said that we are dealing with an issue here that involves *global climate change*, an unforeseen consequence of human economic presumption on a scale of consequence that staggers—that should stagger—the imagination. Our universal response should be in the realm of spiritual shame and intense political engagement. We should be striving to minimize the impact and if possible reverse and undo it: to atone for our sin, to put it in standard religious language. The reality is (no news to anyone) that we are, collectively, in a mode of response that wobbles between denial and bored indifference. We really don't care.

Why don't we care? We do not want to reduce our energy consumption. We do not want to make do with less. We were all raised in the atmosphere of Progress, the momentum of More, Bigger, Better, Faster, and the contrary term for the opposite of all this is "backwardness," with which we want absolutely no association. As former U. S. Senator Theodore Bilbo of Mississippi said of atomic war and miscegenation (in Robert Caro's massive third volume of Lyndon Johnson's biography, *Master of the Senate*), it is "better to see civilization 'blotted out with the atomic bomb . . . than to see it slowly destroyed in the maelstrom of miscegenation, interbreeding, intermarriage, and mongrelization.'"[1] Atomic war is preferable to integration. If a U. S. Senator could (and did) say such a thing—an adjective like "criminal" is entirely too puny for such an utterance—it should, in some ways, come as no surprise that, collectively, we would rather race toward certain disaster than face humbly and bravely into our addictions, distortions, and sins.

The ecological critique is a fine path to wander down, for it offers us a fairly complex view of our suspension. We can see very quickly that we are suspended in our affluence. But that's not all. We are also suspended in our craven political ideology with its myth of achieved moral goodness. Consider this: all the major revolutions of the last 225 years—the American, the French, the Russian, the Chinese, the Cuban (just to name the more obvious and familiar)—were political upheavals that strove to eject a ruling aristocracy and replace those respective aristocracies with a people's or democratic governance that aimed to redirect the very process of government from wealth concentration to wealth dispersal, from elite privilege to common equality. That is our common understanding. In fact, our American experience has seen the formation of a new "aristocracy" of industrial wealth veiled behind huge monopoly corporations with global reach, corporations that are the "private" side of the revolving door with "public" governance. When corporations in the nineteenth century became legal "persons," aristocracy was slyly resurrected in the brand-name guise of commercial identity. Critics like Kevin Phillips, Ralph Nader, and Frances Moore Lappe have demonstrated that wealth concentration is now essentially identical in its proportional imbalance as it was when explicit hereditary aristocracies ruled the peasantry.

We have, accordingly, not only been indoctrinated into the mythology of Progress (with its evil shadow of backwardness), we have also been indoctrinated into the ideology of capitalism (with its evil shadow of socialism); that is, not only does "backwardness" terrify us, we are viscerally repulsed by any program for systemic sharing. So what are the economic and political requirements imposed on us by global warming and climate change? Well, the economic requirement is obvious: we must hugely reduce our consumption, our waste, our pollution. Our throwaway affluence must end. But the politics is more ambiguous. One path—and it is the path we are blindly headed down—leads to a restoration of explicit hereditary aristocracy (even if hidden behind robotic corporations) that lives in elite affluence, fully protected by a swift and very vicious military, and an increasingly impoverished and destitute "peasantry," largely devoid of the cultural aptitudes of the traditional peasantry and therefore more demoralized. The other path requires a political awakening, a suspension of prevailing mythology. It says—Wait a minute! Number One, we have to make do with less while we critically examine our doctrine of Progress. (That's the wake-up call.) Number Two, if we have to make do with less as we sift and sort through our technologies, let's make sure we do this in as full a process of democratic sharing as possible. If less is what's required, socialism is back on the table. In fact, let's just combine the obvious and call it what it is: democratic ecological socialism.

Let's have publicly-owned mass transit. Let's reduce our reliance on private automobiles and trucks to the fullest possible extent. Let's grow broccoli in Wisconsin rather than ship it two thousand miles from California. Let's reinvigorate the culture of the countryside with thousands upon thousands of land-trust gardening and farming communities. Let's see a proliferation of greenhouses, solar panels and wind generators. Let's live here on Earth—on the real Earth, not some virtual earth saturated with technological and religious fantasies of arrogant immortality "protected" by ecocidal weapons.

III

In "A Good Farmer," novelist Barbara Kingsolver says "Most of our populace and all our leaders are participating in a mass hallucinatory fantasy . . . ."[2] "Mass hallucinatory fantasy" is pretty strong language; but it is, I believe, only another way of talking about cultural suspension. Or, to put the two together, what we're suspended in is a "mass hallucinatory fantasy." It's both interesting and relevant that Ms. Kingsolver opened her essay by saying she got deeply interested in the study of agriculture sometime around her fortieth birthday, but told almost no one about it because she feared people would think she was "out of my mind."

Agriculture is, in fact, the bottom of the barrel when it comes to subjects that, as the kids say, are *bor-ing*. We love to eat, to talk about food; but let's not, for Pete's sake, get off on dirt or on how our delectables are raised in dirt. Thank God for agribusiness! For the extermination of small-scale farming (not to mention the liquidation of the peasantry) has enabled us to browse the Internet comfortably in suburban affluence. Nothing could be farther from our minds than that a reconstruction of localized, small-scale, unpoisoned food production is at the very center of ecological, cultural, economic, and political necessities. I would even say that agriculture is at the center of our suspension. Why I think this can be explained rather briefly.

Those scholars and authors (like the late Lewis Mumford) who have dealt at great length with the emergence of civilization, point out that civilization, as an elite, coercive, class-based system of economic exploitation, needed a pre-existing productive agriculture to exploit. This exploitation created a two-class society—an aristocracy and a peasantry. The aristocracy lived in luxury. The peasantry lived in enforced destitution. This pattern persisted openly (despite variations in slavery, serfdom, and generalized peonage) until the industrial revolution (plus very limited democratic revolutions) provided those forms of technology (and a rapidly enlarging consumer market) that enabled the industrializing forces to obliterate the peasantry and (more slowly) undermine small-scale farming. The affluence of the aristocracy with its "civilized" refinements provided the imagery, the ideal, and the goal for society as a whole. What an enormous irony, paradox, and contradiction that at precisely the point where democratic aspirations kick in, historically, the great agrarian base of society is economically crushed and culturally demeaned. "Democracy" implies a total merging with "civilized values," and hardly anyone recognizes what an impossible contradiction this presents.

"Democracy" means that now we can all be—or, at least, imagine ourselves as—aristocrats. Consumerism is based on and allied with precisely this delusion, this "mass hallucinatory fantasy." This fantasy/delusion results directly in the things Barbara Kingsolver itemizes: the waste that poisons, the emissions that are altering climate, the fish and fuels that are depleting, the human deaths as the price of our fiercely blind systemic greed. In other words, we are suspended in the mass hallucinatory fantasy of civilization, in those elite prerogatives and (Mumford's term) "traumatic institutions" that *have always been* the functional core of civilized values. (That we persist in talking about "civilized values" as the repository of common decency, attentive discourse, and polite debate is part of the mythological fog that prevents us from recognizing the utter craziness of our normality.)

Civilization has won and continues to win because its traditional hallucinatory fantasy became, with the industrial revolution and a dollop of democracy, a *mass* hallucinatory fantasy. But since "civilization" is our universal ideal, the name of our secular religion that has lifted us from "primitivity," "backwardness," "paganism," "underdevelopment," and any number of other dreadful negativities that lurk in nature and slither in the past (especially the agrarian past), we simply cannot bear to face and suffer through the recognition of our ideal for what it actually is: an interlocking set of governing institutions based on male violence, elite greed, and relentless exploitation—all of it in pursuit of some sort of egotistical immortality. It *is* scary to face the task of creating a society and a culture without the overbearing directives of civilization—scary *and* exhilarating. But what's scarier is the realization that civilization in full global control is leading us to industrial-military mayhem on a scale of devastation from which we may never fully recover, and that we all are, in some ways, to some degree, caught up in the mass hallucinatory fantasy that carries this catastrophe forward.

We are all complicit. That is a cultural and social fact. Our first task—as deeply and as fully as our limited, flawed consciousness allows—is the recognition of this complicity. Then, unless we opt for who-gives-a-damn ecopathic cynicism, we have to figure out how to climb down from our suspension. I suspect the near future will provide many painful goads to prod us down the ladder.

NOTES

1. Caro, *Master*, 690.
2. Kingsolver, "Good," 13.

7

A Green Critique of Socialist Agriculture

IN THE OPENING ESSAY, I utilized the contributions of the late E. F. Schumacher as a way of exploring the intellectually provocative and culturally revolutionary insight of Lewis Mumford. That insight—civilization as an inherently utopian project—provides us with an exceptionally flexible context by which to assess the explicit purposes, means, and goals of secular and religious ideologies. The view from the garden begins to take on rather extraordinary intellectual and spiritual dimensions.

Green politics is garden politics. This assertion seems fundamentally true even though Green politics is presently constrained by several powerful factors. The first of these constraining factors is the actual political and economic condition of the world in which Green *politics* attempts to act with clarity and effectiveness: prevailing conditions force Green activists to think much more in a reformist mode than they would prefer. Second, Green *culture* is thin, at best, and this means that Green assertions and Green policies (e.g., ecological society demands reduced consumption) make conventional people uneasy: there is not much of an alternative cultural model available that conventional people could themselves utilize as a guideline for personal change. Third, even Green intellectuals and Green activists, as children of utopia, carry an unconscious disregard for rural issues: the major exception is the fascination for "wilderness"; but, as Henry Nash Smith has shown us in *Virgin Land*, "wilderness" is the alter ego of the civilized mind.[1]

In other words, Green politics has yet to exorcize its own urban bias. A small exploration into the intellectual fields of *socialist* agriculture may help clarify the issue for Green-oriented people. By far the greatest intellectual energy of contemporary socialist thinkers has been directed toward conceptions that are primarily urban and industrial. Michael Harrington, for instance, in a section of his *Toward a Democratic Left* dealing with the peasant base of the Chinese revolution, says that "peasants may well be capable of greater political struggle than Marx imagined, but in the ultimate analysis they can only deliver power to the city dwellers who will rule over them." Harrington goes on to say that "Scientific technology is urban, and one could almost define a modern nation by saying it has succeeded in shifting masses from the fields to the factories and, eventually, to the computers."[2] It is necessary in these circumstances to ask two questions. First, what is the nature of the socialist vision? Second, what is the place of agriculture and rural life in that vision?

II

Despite the obvious truth that socialism is built on visions of how things *might* be, a great many socialist thinkers seem tremendously wary of expressing themselves in terms that could be construed as eutopian. Their vision of a fuller and healthier world gets pinched into manageable word-commodities: equity, egalitarianism, social justice, economic democracy, and the like. Unquestionably, these are all fine and important concepts (which Green politics recognizes and affirms), but these concepts don't necessarily add up to a *vision,* and it is *vision* that energizes people toward ecological living and social transformation. In *Unto This Last,* John Ruskin expressed the dynamics of the situation with his usual lucidity: "Three-fourths of the demands existing in the world are romantic; founded on visions, idealisms, hopes and affection; and the regulation of the purse is, in its essence, regulation of the imagination and the heart."[3] We regulate the power of our imaginations, in other words, by the coin of our speech. And when it comes to current intellectual expenditure, we are pretty well-burdened with an overabundance of small change.

Surely the assertion that vision is what people deeply desire and need is borne out by the continued interest in thinkers like Henry David Thoreau or Ruskin himself. Or, to use a more contemporary illustration, we could consider the powerful influence that Schumacher's *Small is Beautiful* has had on the intellectually young. Clear, passionate prose, largely devoid of technical jargon, pulls people out of their normative indifference and enables them to grasp alternatives to the existing order, beyond mere reformist correctives. The enduring influence of such writers as Thoreau, Ruskin, William Morris, Peter Kropotkin, R. H. Tawney, Paul Goodman, and E. F. Schumacher is not due merely to the beauty of their prose, but to the tone and quality of their message. And their message, in marked contrast to writers like Harrington, Robert Heilbroner, John Kenneth Galbraith, or even Barry Commoner, is explicitly more attuned to and in strong sympathy with agrarian considerations. Harrington may approve of the shift of the masses from fields to factories to computers, but R. H. Tawney would have cried out "Computing *what*?" Heilbroner may outline the steps toward achieving the "great ascent" of industrialization, but William Morris would have wanted to know the purpose of the product and whether it had any real *craft* in it. Galbraith may worry about the excesses of the managerial state, but E. F. Schumacher would have pleaded for intermediate technology. Commoner may be genuinely concerned about the effects of agribusiness chemicals on our health, but Paul Goodman would have demanded a thoroughgoing rural reconstruction.

Whether we get an agrarian and decentralist vision from socialists seems to depend on two things. First, whether socialists are providing much in the realm of vision at all, at least beyond the standard "vision" of maximized industrial productivity, and, second, whether their visions embrace agrarian reconstruction. Typically, the essence of socialist society is seen to be centered permanently in urban environments. (Using Mumford's analysis, one could easily say that socialism continues to be saturated with utopia even though socialists may rigorously deny that they are "utopians.") Socialism may be an urban ideology, but the bulk of the world's people are not only poor but rural, although destruction of rural culture has resulted in massive rural-to-urban migration. Cities the world over are teeming with people, millions of whom live in wretched conditions, bereft of hope. Fully granting that class expropriation is a major factor in the creation of alienated poverty, socialism everywhere continues to cherish the utopian ideal of urban life. But *more* urbanization is clearly and unmistakably absurd: we will never again have livable cities until the economic forces and utopian ideologies that mangle, squeeze, and mutilate rural society are recognized and overcome.

It is also true that industrial nations themselves have reached a point in their own development where the costs of resource depletion, pollution, environmentally related diseases, inflation, unemployment, alienation, anomie, and cultural deterioration have become much too massive to ignore or sweep under the synthetic carpet. Global warming has become a household word. Western-style affluence is not only unexportable to the poor and poorest nations, it has become unaffordable in the rich nations, too. All this leads us to a very important question: given that our Western standard of living *in material terms* is going to drop, and given that there are a great many of us wishing to live productive and meaningful lives, what are we going to *do* with ourselves? How, in other words, can we restructure our society and create the conditions for healthier, happier, and more cooperative living even as the material standard declines? And, finally, what is the nature of those conditions and what is the *vision* of that society?

If we will witness a contraction of such industrial commodities as gas-powered automobiles fueled by Middle Eastern oil or electrically-operated air conditioners run off nuclear power plants, we shall also see the dismantling of "export platforms" for foodstuffs coming to the First World from Third World countries—people who can hardly afford to be providing *us* with *their* sorely needed food. We shall also observe the breakdown of the "vertical integration" of multinational agritechnology, and the arrest of undiversified, high-energy and food-poisoning agribusiness in this country. For those thinkers (including socialist thinkers) who see no farther than the three-digit interstate loops surrounding the metropolis of their choice, such contractions (even if ideologically "correct") can only provoke intense anxiety. While the impending contraction of Western affluence is indeed serious, perhaps even grim if we allow it to catch us without systematic correctives underway, it *is* inevitable. It should be clear to us by now, whether by ecological analysis or by the study of past civilizations, that we can manipulate and "master" nature only within boundaries of ecological stability, and we have already gone far beyond the limits, not only of environmental tolerance, but of human oppression as well. Only a Green socialism will be able to get us out of our industrial mess. But a socialism that is not Green will only rearrange the debris.

III

Socialism has had an addiction to organizational permanence and an infatuation with the utopian machine. This is why the most important "class struggle" of the last centuries receives so little attention: the destruction of the rural people, the peasantry. It is interesting to note that the most massive burst of land enclosure in England—seven million acres in the period between 1760 and 1843, according to Arnold Toynbee in *The Industrial Revolution*—occurred *simultaneously* with the emergence of the factory system.[4] Peasants were being systematically driven off the land and the commons were enclosed. Eventually, as industrialists overpowered the landed aristocracy in Parliament, the prohibition on the importation of cheap grain was cast down, and the wages of the laboring poor, already tied to the price of domestic grain, suffered an additional blow.

The calloused attitude of the governing class toward the poor, dovetailing with the views of Malthus on the inevitability of excessive population growth and dwindling food supply, gave to the conditions of poverty a certain aura of geological permanence. At the same time, the living conditions in the new industrial towns in England were wretched. Unionization of industrial workers eventually resulted in an improvement in the living conditions of the working class; but rural society was rarely treated by radicals and reformers with anything resembling primary interest. Few gave much thought to the depopulated countryside or to the urban-industrial policies that guaranteed a forced migration from countryside to city. The shape of the future, whether socialist or capitalist, was explicitly urban.

Socialist ideology postulated the development of urban-industrial abundance as the necessary and logical precursor to socialist society. And for those countries like the Soviet Union, which hurried down the road to industrial expansion, the accumulation of capital was of critical importance. Stalin's reputation for cruel and oppressive ruthlessness is due, above all else, to the social costs of capital extraction from agriculture for massive industrial growth. This passage in Harrison E. Salisbury's *Russia* is worth quoting at length:

> The biggest problems in the Soviet economy almost from the start have centered on agriculture. Many critics trace the difficulty straight back to Karl Marx, who was a city man. His interests lay in the factory system and the urban proletariat. He paid little attention to the European peasant and farming problems. Many scholars feel he had no basic understanding of these matters. His supporters, particularly in Russia, were largely intellectuals or city workers. Lenin's only contact with the peasantry and farm life came during a year or two of his adolescence spent on a family estate in the Volga region. Trotsky came from the city. Stalin was born in a village in Georgia, in the Caucasus, but his whole adult life was spent in the city.

> Lenin at the time of his death in 1924 left farming in Russia almost entirely in the hands of individual peasants. The state operated a few large grain and cattle farms (most of them estates which had been expropriated). There were a handful of cooperative, or "collective," farms. That was all.
>
> Stalin decided to change all this. He had three objectives. First, he wished to break the political power of rural Russia, which could, by simply reducing sown crop areas or by withholding food from the market, bring enormous pressure to bear on the Government. Second, he wanted to extract from the peasants every possible kopek of profit from their production—profit to be utilized in financing industrialization. Third, he thought he could increase farm output by combining the individual peasant plots into larger, better-managed units.
>
> Stalin won his drive to collectivize Soviet agriculture. In a space of less than three years more than 95 per cent of Russian farm lands were incorporated in farm collectives or state farms which were directly operated by the Government. But the price was colossal. Millions of kulaks, or rich peasants, were uprooted and shipped to Siberia. Civil war broke out in some regions. Peasants retaliated against the Kremlin by slaughtering their cattle, burning the harvests, concealing crops from the grain collectors. The toll in lives reached into the millions and many regions were struck by famine. The wounds to agricultural production were so severe that more than 30 years later they are still to be felt . . . .
>
> But two of Stalin's objectives were fulfilled. He did break the political power of the peasants and he did squeeze out of them the funds which he poured into the gigantic industrialization program.[5]

Salisbury doesn't quite tell us, however, where the "funds" were obtained. In an article called "Ukraine" in the May 1987 *National Geographic*, writer Mike Edwards provides us with another piece of Stalin's puzzle:

> Having forced the peasants into collective farms despite their stubborn resistance, he imposed impossible production quotas. Police and party brigades bore off everything edible. While the famine raged, the Soviet Union was selling grain abroad. More than half a million tons were exported in 1932, and in 1933 shipments even increased by a third.[6]

What makes the Soviet experience unique is not the expropriation of agricultural commodities but the extraordinarily rapid transition from an essentially preindustrial pattern of expropriation, which left the subsistence culture of the village more or less intact, to a kind of expropriation that required a utopian reorganization of rural life. This is the process that Edward Crankshaw, in *Khrushchev's Russia*, describes as the transformation of "dark villages with their stubborn traditions and superstitions" into a "clean well-regulated, antiseptic paradise."[7]

Robert Heilbroner, in *The Great Ascent*, describes in exactingly rational detail the economic and political steps in the "engineering of development." One of the key steps in the process of industrialization (and "development" is synonymous with industrialization in the technospeak of contemporary utopians) is the forcing of the smallest agriculturalists off the land and into the "labor market," and the "formation of land units large enough to be farmed scientifically for high output."[8] The countryside, in other words, is to prime the industrial pump, but the commodities will flow into the city. The industrial revolution succeeded in England and the Soviet Union largely because the agricultural economy and agrarian patterns of life, as the economists like to say, got squeezed. All industrial economies seem to follow the same pattern. (From ninety-five percent in 1790, the agrarian portion of the American population now dips well below two percent.) And it's not as if the industrial economy, once successful, relaxes its grip on the countryside. Not at all. The squeeze stays on. *The squeeze becomes institutionalized*. And a "cheap-food" policy is only one aspect of the undervaluation of "raw materials" in an industrial utopia.

Given that the economic forces of industrialism compel the contraction of agrarian society and force into being "land units large enough to be farmed scientifically for high output," we still have to understand the ecological and social costs of such a major transformation. One rather stark way of grasping the dynamics is through seeing that "there is no more folk art because there is no longer anything that could be called 'folk'" (Arnold Hauser, *The Philosophy of Art History*)[9] or by realizing that "increasing industrialization irrevocably destroys folk art. The possibility of folk art now renewing itself by drawing upon the content and means of expression of the peasantry and the wandering artisans has become extremely remote" (Ernst Fischer, *The Necessity of Art: A Marxist Approach*).[10] It is more than a little ironic that socialism, in its carrying out of policies in the name of the common people, has contributed immeasurably to the destruction of common culture, just as "progress" in the early decades of

England's industrial revolution meant material impoverishment and cultural mayhem for the bulk of the laboring poor.

While it is unquestionably true that material conditions have vastly improved for the majority of people within the industrialized nations—while conditions have probably worsened for the majority of those in the Third World—it is also true that the appreciation for small, self-regulated communities, so strongly advocated by such divergent thinkers as Thomas Jefferson and Martin Buber, has become a utopian joke. But the rise of social *anomie*, in both East and West, should inform us that the spread of industrial affluence has not given us the high road to better human community—not with thirty thousand atomic warheads as "peace keepers." We have shrugged off the feudal and monarchic authoritarianism of the past only to have moved into an urban-industrial "future shock" of deepening atomization.

In ecological terms, we can see that the costs and effects of "scientific" farming are of a piece with industrial pollution generally. The urban demand for cheap food, in concert with the development of technological methods for greatly increasing yields (which enable the agribusinessman through mechanical and chemical processes to control ever-larger acreages), has introduced into agriculture some deadly practices and side effects. The data are already coming in that farmers who use chemical herbicides and pesticides suffer a higher-than-normal cancer rate. Groundwater is polluted. Giant machines compact the soil and use more caloric-quantities of energy in the form of fossil fuels especially than the quantities of food-calories produced. Undiversified mono-cropping (even the alternating of soy beans and corn) destroys natural fertility and depletes humus content in soil. Soil erodes by wind and water, bakes under the sun, in fields barren of cover crops. Chicken factories, hog and cattle feedlots, are not only a brutal reductionism of animal "husbandry," they are also serious pollution problems. And in the ecology of esthetics, the rapid proliferation of metal buildings, massive machines, mercury vapor lights in farm yards, great concentrations of animals who are never permitted to wander or graze, and the omnipresent agri-industry caps (brand names over the visors) on the heads of agribusinessmen, all point to agrarian sterility. Stewardship, cultivation, and husbandry have all suffered heavily under the imposed instrumentalities of industrialized agriculture. Agri*culture*, in point of fact, has been pushed aside by expansive agri*busi-*

ness. The commons were enclosed. The peasants were evicted. The folk have been successfully displaced.

IV

Shepard B. Clough, in *The Rise and Fall of Civilization*, identifies two periods when the "development of new techniques was crucial in economic progress." The first was "the millennium prior to 3000 B.C.," and the second "in the four and a half centuries since 1500."[11] The first period corresponds to the rise of civilization in the area sometimes known as the Fertile Crescent. The second period corresponds precisely to the rise of imperialist civilization, centered in European nations, on a scale unprecedented in world history. It is also interesting to note that the plow has been traced back to approximately 3000 B.C. This has particular meaning in light of the feminine origins of horticulture. V. Gordon Childe, in *What Happened in History*, insists that the introduction of the plow "relieved women of the most exacting drudgery, but deprived them of their monopoly over the cereal crops and the social status that conferred."[12] In *The City in History*, Lewis Mumford says that the "most important agent in effecting the change from a decentralized village economy to a highly organized urban economy was . . . the institution of Kingship," and that this change was "accompanied by a collective shift from the rites of fertility to the wider cult of physical power."[13]

There are many social, political, and economic features of the early urban transformation, but let us consider two. On the one hand, men took over some agricultural practices from women (hunting having become largely obsolete), and, on the other, the peasantry *as a whole* was made increasingly subservient to urban power. Technically, this latter point meant that the city (through taxation and armed force) was extracting capital from the countryside to finance programs that, in the main, enhanced urban power and prestige. But in the second period that Clough identifies as critical in economic progress, we find the city extracting capital from the countryside on a scale never before possible in world history—a scale of exploitation made possible by the vast array of technical advances associated with the rise of industry. In other words, the emergence of the city contracted the feminine sphere of agriculture—while replacing the Goddess with the God—and gave rise to an exploited peasantry: civilization's packhorse, as Trotsky supposedly called the peasants. But the second burst of urban power *destroyed the peasantry itself* and "gave birth" to an instrumental, soil-destroying agri-technology. Until the modern period, the city always had to content itself with *domination* of the peasantry; but now the city has concentrated the power by which to *obliterate* the rustics altogether.

Socialists, insofar as they advance the cause of urban industrialism at the expense of agrarian society and the living soil ("rising above backwardness," "conquering nature"), partake of the traditional, elite, utopian mentality. If the elite class system, endemic to the historical city, is the focus of socialist outrage, then the exploitative urban policies of many modern socialist politicians (and of socialist political theory in general) is an amazing commentary on the power of ideological distortion. If the socialist claim is a true one—that capitalist economies are based on exploitation and inequality—then socialists need to look two historical realities straight in the eyes: that civilization as a social structure is *constructed* of class stratification, and that civilization as an economic structure requires the *enforced* extraction of "surplus value." Industrial civilization in a socialist form is a contradiction in terms. A nonexploitative and egalitarian economic order has to be ecological in a social as well as in a natural sense. That is to say, the prerequisite to ecological well-being at the level of agriculture and general land use is a symbiotic ecology between countryside and town. Such a symbiosis requires the contraction of the urban-industrial ethos, of utopia; it requires the development of deliberate policies to revitalize and reconstruct rural culture; it requires a far greater emphasis on the *quality* of living, both culturally and ecologically, as opposed to the mere materialist *standard* of living.

A real eutopian "post-industrial" society can only be sustained after the raw aggressiveness of city-based nation-state systems has reached its obsolescence and is headed into real decline. Yet civilization is conventionally understood as a process of *spiritual* domestication, with the city as its leading organ. The decline of the city, the contraction of civilization, therefore seems to suggest the abandonment of a certain valuable spiritual process, the decline of the "civilizing tendency" itself. Let's briefly examine this idea by returning to an analysis of the city's origins.

V

In her book *On Understanding Women*, historian Mary Beard contends that women, besides domesticating grain, were first to domesticate animals, control fire, practice medicine, spin and weave, make pottery, sew clothes, and build shelters.[14] A great many other authorities agree in general with this assessment, although some would give the domesticating of animals, shelter building, and the controlling of fire over to the initiative of men. The contributions that women made toward the development of cultural forms leading to the creation of the city cannot be stressed strongly enough. Beard asks us to "see primitive women as the inventors of the domestic arts—cooking, spinning, weaving, garnering, guarding, doctoring, providing comforts and conveniences, and making beginnings in the decorative arts. They are launching civilization."[15]

Historian Gerda Lerner, in *The Creation of Patriarchy*, provides a synopsis of Elise Boulding's "overview of women's past," as condensed from the latter's *The Underside of History: A View of Women Through Time*:

> Boulding sees in the Neolithic societies an egalitarian sharing of work, in which each sex developed appropriate skills and knowledge essential for group survival. She tells us that food gathering demanded elaborate knowledge of the ecology, of plants and trees and roots, their properties as food and medicine. She describes primitive woman as guardian of the domestic fire, as the inventor of clay and woven vessels, by means of which the tribe's surpluses could be saved for lean times. She describes woman as having elicited from plants and trees and fruits the secrets of transforming their products into healing substances, into dyes and hemp and yarn and clothing. Woman knew how to transform the raw materials and dead animals into nurturing products. Her skills must have been as manifold as those of man and certainly as essential. Her knowledge was perhaps greater or at least as great as his; it is easy to imagine that it would have seemed to her quite sufficient. In the development of ritual and rites, of music and dance and poetry, she had as much of a part as he did. And yet she must have known herself responsible for life-giving and nurturance. Woman, in precivilized society, must have been man's equal and may well have felt herself to be his superior.[16]

In all this, Boulding's analysis is very close to Beard's and Mumford's. Lerner goes on to probe for important changes in Neolithic society:

> Approximately at the time when hunting/gathering or horticulture gives way to agriculture, kinship arrangements tend to shift from matriliny to patriliny, and private property develops. There is . . . disagreement about the sequence of events. Engels and those

> who follow him think that private property developed first, *causing* "the world historic overthrow of the female sex." Levi-Strauss and Claude Meillassoux believe that it is the exchange of women through which private property is eventually created.[17]

These changes in kinship arrangement lead Lerner to insist that "in the course of the agricultural revolution the exploitation of human labor and the sexual exploitation of women become inextricably linked":

> There are a few facts of which we can be certain on the basis of archaeological evidence. Sometime during the agricultural revolution relatively egalitarian societies with a sexual division of labor based on biological necessity gave way to more highly structured societies in which both private property and the exchange of women based on incest taboos and exogamy were common. The earlier societies were often matrilineal and matrilocal, while the latter surviving societies were predominantly patrilineal and patrilocal . . . . The more complex societies featured a division of labor no longer based only on biological distinctions, but also on hierarchy and the power of some men over other men and all women. A number of scholars have concluded that the shift here described coincides with the formation of archaic states.[18]

By "archaic states" Lerner clearly means early civilization. She says the transition from agricultural villages to towns to cities to "archaic states" is a

> . . . process which occurs at different times in different places thoughout the world: first, in the great river and coastal valleys of China, Mesopotamia, Egypt, India, and Mesoamerica; later in Africa, Northern Europe, and Malaysia. Archaic states are everywhere characterized by the emergence of property classes and hierarchies; commodity production with a high degree of specialization and organized trade over distant regions; urbanism; the emergence and consolidation of military elites; kingship; the institutionalization of slavery; a transition from kin dominance to patriarchal families as the chief mode of distributing goods and power.[19]

This, needless to say, is the very process that Mumford identifies as utopian.

The problem with Lerner's analysis is the implication that agriculture *is to blame* for the transformation from "relatively egalitarian societies" with a sexual division of labor to "more highly structured societies" with private property and the subjugation of women. The new capacity to produce food in greater abundance cannot itself be blamed for the systematic expropriation of that very abundance. This is the familiar tendency to blame the victim for the crime. To be sure, agricultural fecundity was open to external control; but *agriculture did not expropriate itself.*

VI

If women claim oppression by men since at least the emergence of the city, certainly the peasantry as a social class can assert a parallel claim against civilization as a whole. Just as in the sphere of the individual psyche, where Freud has shown that whatever is repressed does not necessarily disappear, so agriculture has retained a historical suspicion of the city. The very foundation of Freud's thought, in the words of Norman O. Brown's *Life Against Death*, was "based upon the theory of repression."[20] And repression is a real connecting link between women's history and the history of agriculture: "the power of some men over other men and all women."

Freud failed to develop an understanding of the peculiar nature of men's repression of women. Marx failed to develop a sympathetic understanding of the peasantry's oppression by civilization. The connecting link between women's repression and agriculture's oppression is, therefore, male-dominated utopian civility. Engels postulated a precivilized period of "primitive communism" under the rule of women—"matriarchy." Freud reached into the unconscious and discovered the pre-Oedipal mother. But neither of these men ever came to the conclusion that the amelioration of rural oppression and women's repression is keyed to the restoration of women's dignity *and* the recovery of the eutopian garden. In other words, neither Marxism nor Freudianism adequately addressed the oppression of women, the peasantry, and nature as the mode of civilization itself.

The women's movement may well be the single most important social phenomenon of our period. On its success rests the possibility of eutopia, ecological society, and the survival of humanity—provided women see through the inherent contradictions in the predatory male design of utopian civility. The utopian military machine is, obviously, governed in its actual functioning by men. The loosening of male governance implies the loosening of utopian rigidity and civilized exploitation—provided, again, that women fully recognize the utopian function of the industrial machine. So far it is primarily eco-feminism (that relatively small grouping of feminists who have discovered powerful linkages between women's history and civilization's treatment of nature) that has begun to address the social ecology of the countryside. But in challenging the rapaciousness of the civilized machine, eco-feminists call into question—implicitly and explicitly—the utopian "standard of living." This, in turn, challenges the normative lifestyle within industrial society, which in its turn activates deeply-seated anxieties about the "backwardness" of ecological living in the countryside. The ecological crisis—namely, that the Earth cannot support a utopian standard of living—will serve even more than it has already to bring pressure on the dilemmas of gender.

Yet the urban bias against rural life that runs so deeply in civilized consciousness—deeply enough to be simply taken for granted, therefore normative and "unconscious"—is a critical issue that needs deep attention from all who wish to undo the military-industrial complex and nature-shock technologies. Civilization rests on exploitation and expropriation. The select dismantling of the utopian machine must be balanced with the creation of a eutopian folk culture. The economy of human society must be re-embedded in nature's ecology.

NOTES

1. Smith, *Virgin*, 293.
2. Harrington, *Toward*, 211.
3. Ruskin, *Unto*, 78.
4. Toynbee, *Industrial*, 12.
5. Salisbury, *Russia*, 52–53.
6. Edwards, "Ukraine," 594.
7. Crankshaw, *Khrushchev's*, 79.
8. Heilbroner, *Great*, 78.
9. Hauser, *Philosophy*, 330.
10. Fischer, *Necessity*, 67.
11. Clough, *Rise*, 260.
12. Childe, *What*, 81.
13. Mumford, *City*, 35.
14. Beard, *Understanding*, 41.
15. Beard, *Understanding*, 514.
16. Lerner, *Creation*, 43.
17. Lerner, *Creation*, 49.
18. Lerner, *Creation*, 52–53.
19. Lerner, *Creation*, 54.
20. Brown, *Life*, 3.

8

The Meaning of Green Agriculture

THE WORD "GREEN," AS applied to gardening, farming, and agriculture, is, in my estimation, pretty poorly understood. In fact, the implications of a Green food supply probably will cause some uneasiness even to those of us who like to think in terms of "organic" or "sustainable" food production. The word Green, in the broadest conception of the *culture* of food production, means a great deal more than just a set of techniques or methodologies applied to food-growing processes. Green means a great deal more than "parity" or fair price.

In this society, we are heavily oriented toward a "how-to" approach, we are inclined to think in terms of profit and reward, and so we are disinclined toward, and even impatient with, an analysis that requires us to think in historical, cultural, and spiritual patterns. So, knowing that you can obtain excellent how-to information from a multitude of sources—how to garden, how to farm, how to spin and weave, how to put together your own alternative energy equipment—I am going to tax your patience by talking about history and culture. First I will remind you of an obvious thing. The biggest word we have for all the aspects of food production is not agri-method or agri-technique or even (though it's become so) agribusiness but agri-*culture*. And the Latin roots of agriculture come from words that refer to field cultivation. The words "cultivate," "culture," and "cult" are all related and they reflect, sequentially, the practical methodologies, the social relations, and the spiritual or religious significance of how we grow, share, and are grateful for the Earth products that enable us to live, individually and collectively.

A good part of the Green criticism of, and even anger toward, agribusiness lies in the recognition that agribusiness represents a kind of how-to, chemically-reductionist tunnel vision—concerned only with maximized yields and maximized profits. ("Get big or get out," as former

Secretary of Agriculture Earl Butz proclaimed.) Agribusiness does not care about, and is not interested in, the cultural meaning and spiritual content of nonmarket food production. The word "market" is a kind of cultural gate between agriculture and agribusiness. That is, if you believe that the only real way to understand food production is in terms of the market, you are by definition in the camp of agribusiness. Now you may argue for an "organic" agribusiness because of concerns about chemicals or genetic engineering. That would put some Jolly Green make-up on the agribusiness giant.[I] We do realize, though, without quite grasping or being able to predict the dimensions, that an "organic" agribusiness would necessitate smaller farms and more labor, for if you can't poison a field or clone critters into submission, you've got to spend more hands-on time out there cultivating. And that means being more labor-intensive.

This is a small step toward what "Green" means and requires—and we haven't yet come close to Wendell Berry's insight that it is "the subsistence part of the agrarian economy that assures its stability and its survival."[1] We still haven't gotten to much in the line of history or culture. So here we go—and I'll give you as short a version of my understanding as possible. Agriculture is relatively new in human history. It originated in the Late Stone Age and—so the anthropologists, archeologists, and historians tell us—derives primarily from gatherers, not hunters. That is, women as the foragers for roots, nuts, fruits, and seeds discovered, and were the first to practice, the intentional planting of crops for human consumption. The abundance so produced over time resulted in a number of things. It led to villages that were stable in location, because people no longer had to follow the food, as it were. It led to a greater density of human population. It made hunting somewhat obsolete because of the eventual domestication of various animals, including the cow, sheep, goat, pig, donkey, chicken, and horse. This abundance has led one great historian, Lewis Mumford, to talk of "the Golden Age" of the agrarian village prior to the rise of civilization proper.

I am now, perhaps, going to challenge your understanding of history as well as your understanding of what Green implies. I want each of you to take a few seconds to weigh and measure, inside yourselves, the ethical and moral valuation of the common words civility, civilize, and civilization. If you grew up in the same conventional cultural atmosphere I did, you will find inside yourselves a positive meaning for these words. And if

I. For a rather thorough study of "The paradox of organic farming in California," see *Agrarian Dreams* by Julie Guthman.

I were to ask you to name the negative opposites of civility, civilize, and civilization, you probably would come up with words like savage, barbarian, villain, heathen, or pagan. I will tell you right now that my cultural understanding, largely unconscious, floated in this conventional terminology until I began to probe, in determined seriousness, for the roots of what we now call the "farm crisis." In the briefest formulation, this is what I discovered: civilization is the perpetual enemy of agriculture; the "farm crisis" is as old as civilization, and I will try to explain exactly what I mean in the shortest possible time.

Every civilization properly identified—from the Babylonian to the Aztec, from the Chinese dynasties to Stalin's Soviet state, from Plato's Greece to Cicero's Rome, from the Egyptian Pharaohs to the American Presidents—has been, through violence, organized in such a way so as to cause wealth to accumulate in the hands of a few. In classical terms, the name for this wealthy few is aristocracy, and in case you haven't been paying attention, we claim to have democratized civilization although we very much continue to have an aristocracy of wealth—and nobody seems particularly puzzled by this obvious and outrageous contradiction. At the root of the aristocracy of wealth lies civilization proper, for civilization is the predatory theft of agrarian abundance from the villages of the "golden age" and from all subsequent peasant villages and farming communities. We have no historical model for a civilization that does not steal, by one method or another, the production of primary producers, and at the base of primary production stands the peasant village, for nothing is more basic than our need for food.

With the emergence of civilization came institutionalized warfare and institutionalized slavery. These two realities—the expansion of the boundaries of empire through war and conquest, and the enslavement of vast numbers of conquered people who were forced to produce and construct—lie open to view in the history of civilization up until the implementation of the industrial revolution. At that point, while expansion of empire in some instances wildly accelerated, the overt enslavement of people was mitigated by new technologies that now enslaved nature directly, with machines and chemicals, rather than with human labor. Well, you may say, even if all this historical claptrap and cultural gobbledygook is true, so what? What does any of that have to do with the meaning and politics of Green agriculture?

Needless to say, Green politics has emerged out of the growing ecological crises and growing ecological awareness of the last sixty years or so—since Hiroshima and Nagasaki, say, and one of the main ecological insights (regarding global warming, for instance) has to do with the relatedness of so many seemingly separate aspects of economic life and human conduct. Another Green insight is this: the total commercialization of food production—the destruction of what Wendell Berry calls "the subsistence part"—occurs precisely as civilization achieves a historically unprecedented degree of global control. The *culture* of the peasant village, of small-farm neighborhoods, is in the process of being destroyed all over the world. The self-provisioning aspects of noncivilized food production are being globally wiped out, and the murdering thief who overran the agrarian villages at the end of the Late Stone Age is now in command, with his terminator seeds, the world over.

What this means is that Green politics is a revolutionary force, for at the core of Green politics is Green agriculture, and at the core of Green agriculture is the unconditional demand for the rectification of the great historic evil committed by civilization against the agrarian village and the Earth. The ultimate goal of Green politics is to encourage and create the conditions for ecological living on Earth. At the core of ecological living is the question of the culture of our food production, for our most fundamental need is to eat, and we eat from life that grows on Earth.

The most basic ethical teachings of the world's great religions stand on two legs. One leg is stewardship in Creation. The other leg is sharing among our own kind. Green politics embodies a new synthesis of decentralized, ecological, democratic socialism integrated with a compassionate, Earth-based global stewardship. Green politics does not yet recognize how politically revolutionary and ecologically conservative it truly is. The eventual victory of Green politics requires the cooperative resettlement of the countryside, an economy of real needs, energy conservation, demilitarization, reverence for nature, racial integration, gender reconciliation, spiritual respect—and the end of civilization as we know it. Green thinking has begun to lead us out of the civilized slavery of consumerism. But we have to do our part. We not only have to envision Green culture, we also have to trust that Green culture is sustainable, and that living it will be a blessing and a joy.

Green culture is, in fact, the only sane option, for civilization supreme is violence, murder, theft, and slavery supreme; and that kind of supremacy, in a world with literally thousands of nuclear missiles, warheads, and bombs, points toward global disaster, a mammalian extermination waiting to happen. The Green political task is multidimensional, as the Ralph Nader and David Cobb candidacies have shown. There are a few brilliant leaders who, by their selfless dedication to issues of public policy, have an extraordinary grasp not only on the complexity of ecological, cultural, social, political, and economic problems, but a set of powerful proposals for structural correctives. Although Green politics includes basic socialist insights and proposals, Green is *not* about creating a larger and larger economic pie. In this respect, Green politics is, of ethical necessity, daringly unique. More of everything spells ecological disaster; and so Green stands alone in its demand for the disciplining of consumption, for genuine economic restraint, and even for a kind of environmental asceticism.

This, again, is a cultural gate with spiritual hinges. That is, Green *culture* presupposes a certain Creation-based spirituality that is not only willing to forgo the shallow allurements of sensational modernity, it also presupposes a fundamental trust that a good, culturally rich life is not only *possible* in the exigencies of ecological living but, by taking our foundational spiritual teachings seriously, *inevitable*. But because the agrarian village lies, humanly speaking, at the base of the civilized pyramid, squashed by the economic, political, and cultural presumptions of civility, the liberation of the agrarian village—of agriculture—is the keystone of Green revolution. Agriculture is the soul of Green politics: an ecological socialism, a resurrected peasantry, and the culturally rich, cooperative reinhabitation of the Earth.

NOTES

1. Berry, "Back," 16.

9

Victory over Dirt

AGRICULTURE EXISTS BECAUSE GATHERERS developed horticulture—plant domestication—in the late Stone Age. The next step, once plant yields reached a certain critical abundance, was domestication of those animals (cows, sheep, goats, chickens, etc.) that shared a portion of the human diet and that could now be fed out of horticultural abundance. Plant and animal domestication, jointly practiced, is the meaning of agriculture. We also, at a more basic level, have agriculture because we are biological beings whose bodies require the fuel that food provides. So far as I know, all biological beings require such food/fuel. Life, we might say, feeds on life, with the fine roots of plants, surrounded by reciprocating bacteria and other microscopic critters, utilizing the soil as a kind of bottomless geological refrigerator, pulling nutrients into the stem, leaves, flowers, and fruits in a seemingly endless cycle of ashes to ashes, dust to dust, dirt to dirt.

But all life, once incarnate, tends to resist returning to dirt. The readily observable flight-or-fight behaviors of mammals, birds, reptiles, and insects show how fully diffuse and entrenched is the survival impulse in various life forms. We like to live. We don't wish to die. We fervently resist physical decline into illness, disease, and aging, that is our exit ramp into dirt. (Plus, even dead, we humans strive to prevent the body from returning to dirt by means of funeral home delaying tactics and cosmetic gimmicks.) In the current civilized effort to finally overcome and overpower "backwardness" in all its manifestations, we can see a certain utopian breakthrough—namely the rise of civilization itself, from its largely Asian origins, now imposed as an ideology, actually a mythology, on the entire world. The effort to rise above dirt, to rise above and never return to dirt, is a fierce utopian impulse encased in class ascension that ruthlessly strives to achieve immortality or a semblance of immortality.

Civilization, in its classic forms, consists of an armed aristocracy imposing its will and extracting "surplus" from a laboring peasantry. Because literacy and the fine arts, elegant architecture, etc., were claimed by aristocracy as its civilized domain and heritage, civilization as a concept with borderline divinity became the ego ideal, if you will, the governing image, even of societies such as our own, that achieved elementary revolutions in and toward democracy. Democracy was implicitly, if not explicitly, understood as a further development, a modification, of aristocratic civilization: the enabling means by which commoners could enter the restricted sphere of civilized governance and commodity abundance. But here there was, and there remains, a really huge problem. That problem, simply put, is that civilization's origin and history are characterized by a set of governing impulses, images, and institutions that were, and are, thoroughly undemocratic, thoroughly aristocratic, thoroughly elitist, and thoroughly driven by an impulse to rise above dirt, backwardness, and mortality.

So as civilization achieves unabashed global dimensions, mangling if not overtly exterminating indigenous cultures and subsistent agrarian communities, it forces into being utopian agribusiness, utopian agritechnology, that scorns dirt except as an anchoring medium for genetically engineered plants "fertilized" by manufactured chemicals and protected from nature's swarthy locals by herbicidal swat poisons. Meanwhile, domesticated animals also undergo genetic molding, are housed in giant automated feeding facilities, are fed medicated and genetically modified foods, are slaughtered and packaged in state-of-the-art utopian dismemberment facilities before distribution to fast-food franchises and supermarkets, and purchased by consumers living in an unbelievably immense and complex technological utopia.

The lesson I take from all this is an extremely troubling one and, if true, suggests our current global crises are not only interrelated but astonishingly pervasive in our everyday lives and in the very patterns of our mundane thinking. What I mean is this: the naive presumption that *civilization* can be, and has been, democratized needs urgent reconsideration. If civilization in its founding impulses was elitist, predatory, and contemptuous of natural cycles—and I believe the record shows this to be exactly the case—why do we persist in imagining that by a little electoral magic we have succeeded in "democratizing" civilization? If civilization is killing us in its pathological drive to substitute utopian technologies for natural processes, resulting in an array of disasters from species extinctions to climate change (to name only two things), why don't we begin considering the abandonment and abolition of civilization? The very thought, needless to say, is received almost universally with incredulity—from satiric amusement to angry denunciation that, in one respect, shows precisely how deeply attached we are to the concept. At its most traditionally breathless, the prospect of doing without civilization invokes Thomas Hobbes' each against all, primal terror, backwardness in its most primitive and bloodthirsty forms—anarchy in its most depraved, conventional sense.

We have a cultural mythology, explicit and standardized, that instructs us that civilization is the chalice in which all true humanitarian desires are to be gathered and tasted. Everything outside this chalice is primitive, backward, and finally evil. I propose, simply, that this self-serving mythology derives from the immortality lusts within classical aristocracy and, furthermore, that our weddedness to this mythology makes us acquiesce to and complicit in utopian predation. Civilization is *not* the crucible of literacy, the fine arts, exquisite music, delicate furniture, and gourmet cooking. Civilization is the concentration of governing institutions that are controlled by and primarily reward the aristocratic elite.

At the base of civilization's "pyramid" are not the industrial workers—this is socialism's great analytical flaw—but Earth, dirt, clean water, gatherers, hunters, and the agrarian village. Green politics informs us, if we care to meditate our way to the core of this understanding, that to live ecologically on Earth requires not only a reconciliation with mortality, with dirt, but a profound commitment to live with strict, but gentle, humanitarian sharing on a global scale. In a spiritual sense these ethics are grounded in stewardship and servanthood, Green and Red, ecology and socialism. They are at or near the center of the ethical prescriptions of every great spiritual tradition. I suggest that such a cultural configuration, if it could be achieved, would *not* be civilization in any recognizable or traditional sense.

II

In Acts 13:13–52 of the Christian Bible, the Apostle Paul, speaking in a synagogue in Antioch, tells his listeners that Jesus rose from the dead "never to return to decay." It is immortality that Paul propounds to his auditors as the key and critical feature of Jesus' meaning. No mention is made in this text of love, healing, sharing, or stewardship. Jesus is to be "believed in" because he rose from the dead and "did not experience decomposition." (In Romans 8:21, it is "creation itself" that Paul hopes "will be liberated from its enslavement to decay.")

I was raised a mainstream Protestant, left the church (like countless others) in my late teenage years, and only returned to it, in a qualified way, in my mid-forties. My intent in returning was, in part, to more deeply discover and, if I could, promote the "social program" implications of the "kingdom of God" so vividly advocated by Jesus, especially in the first three Gospels, and most of all in Luke. (I found considerable assistance in this effort from Joan Chittister, Marcus Borg, John Dominic Crossan, and—the really bad, bad boy of Christian theology—Thomas J. J. Altizer.)

After several years I left again, out of intense frustration with the "salvation club" atmosphere of the church. (Actually I jumped ship to the Quakers, and I seem to have found there a spiritual lodging.) But my more intensive reading of the *New Testament* led me to the conclusion that the salvation club aspect of Christianity is firmly lodged in the bulk of the "books" that follow the first three Gospels. Now, nowhere did I look for or expect to find a repudiation by Jesus of "salvation" or a hereafter. So it is *not* my argument that Jesus was totally about what we might call the "social gospel." But it *is* my position that the first three Gospels reveal a Jesus overwhelmingly engaged in the unconditionally here-and-now dynamics of a spiritual life based on radical sharing and radical stewardship. These dynamics—radical sharing and radical stewardship—are a huge part of the means by which the "kingdom of God" is made manifest.

Do you believe in God? Very well, your task, like mine, is to love God with all your strength and your neighbor as yourself. There are no lawyerly qualifications to this prescription, except insofar as the injunction to love becomes explicitly inclusive even of, and specifically toward, "enemies." These dynamics deepen toward sharing and stewardship precisely because of the deepened requirement to love. This is not an optional interpretation, but by turning Jesus into an exclusive and strictly supernatural salvation figure—and this happens explicitly from the Gospel of John onward—the "kingdom of God" is reduced to virtual triviality and "believing in" Jesus as the divine magnet of afterlife salvation becomes the primary focus and overwhelming preoccupation of Christian endeavor: *not* the transformation of human personality and culture via the spiritual commitment to radical sharing and radical stewardship, but personal and private assurance of heavenly life after death. (The idea of a "Christian nation" or a "Christian civilization" is merely this privatized dynamic packed into mass pseudospiritual doctrine.) Passionate concern for *this* world will, in this milieu, single you out, negatively, as a person inadequately committed to salvation.

A major idea of such salvation is first anticipating and then achieving a state of being "never to return to decay," that is, becoming immortal. Everything truly spiritual, from this perspective, is in fervent anticipation of a life that only really begins after death, in a place that is not this Earth. What is here on Earth is a kind of mortal holding pen, only to be endured. I suggest that a huge proportion of our intensifying global crises has roots fully entangled in this quest for immortality, this rising above the clutches of decay, this "spiritual" acceleration beyond mortal dirt, this effort to achieve a supposedly foolproof intelligence and enter into the actual mind of God. Civilization, it seems to me, in its inception and origin is, or is related to, an impulse to achieve and sustain immortality and the ongoing blessing of divinity. We can, in fact, openly see this in the supposed "divinity" of god/kings, the elaborate mummification of "divine" corpses, and even—closer to our own time—the "divine right" of kings, that has "democratically" transmogrified into our very own Manifest Destiny.

A salvation religion such as Christianity, with its unabashed doctrinal imperialism (either you "believe in" Jesus or your goose is eternally cooked), therefore becomes the "theological" arm of a civilized continuity (Roman, Greek, Egyptian, etc.) that also contains an unabashed imperialism based on conquest. These impulses have been so explicitly congealed in the George W. Bush administration that it makes one's mind reel. This is far less an official merging of church and state than the alternating current of salvation in implicitly religious and explicitly civilized intensities. Both arms of this imperialism wish to raise the victory sign over dirt.

These imperialisms of immortality make stupid, immensely and deadly dangerous, the men who plan for economic advantage and military conquest. This is true, without the slightest shred of doubt, in the area of nuclear armaments, as analyses by such authorities as Helen Caldicott and Jonathan Schell have made terrifyingly clear. But it is also true with the Navy sonar that kills whales, the genetic modification in plants already spreading into natural botany, toxins in "the environment" deadly for thousands of years, global warming and climate change from greenhouse gasses—to mention only a few of the plagues already with us. These plagues can be manufactured and arrogantly deployed precisely because dirt doesn't matter, because we are on a trajectory of progressive immortality, because the Earth is only a crude and clumsy footstool for the glorious eternity soon to come. I am, I'm afraid, stating this poorly, but we're in the incredible pickle civilization has created—human beings, acting in and through history, governed and directed in their institutions by civilized mythology—very largely because of accrued delusions and obsessions regarding immortality: the ideal, the perfect, and the permanent. These delusions and obsessions are endemic to aristocratic civility. Just pick up any glossy high-end magazine and *look* at the ads oozing perfection and power: they represent aristocratic civility in a capitalist technological mode; they project the drive to never return to decay; they are congealed images of spiritual narcissism totally out of control even as their theme is total control and total mastery.

The outcome of this drive to rise above and never return to decay is genocidal and ecocidal murder. The utopian obsession with control is now unsustainably beyond control precisely because its lethality is ecocidal. It is, to borrow a term from Carolyn Merchant, the death of nature. It is the yearning for immortality seized upon and wielded with superhuman, superrational technological abandon. Its meticulous rationality is a control box of immortal madness determined to hold Earth hostage in its demented claim to achieve utopian immortality. Our world is in the hands of utopian terrorists obsessed with a salvation of the ideal, the perfect, and the permanent. Explicitly religious and explicitly civilized immortal terrorisms have congealed as civilization achieves the utopian triumph of actual globalization. The outcome of this obsession will be global catastrophe. Our only hope is that it will not be an absolute catastrophe.

10

A Landscape Disfigured

There are concepts whose accrued and imputed sanctity is so great that critical thinking about them is virtually prohibited. *Civilization* is such a concept. On this continent, "civilization" was the ideal that justified the crushing of indigenous "savage" cultures. And, if we take civilization as the dominant secular concept, its religious counterpart is, obviously, institutional Christianity, in whose name "pagan" beliefs and practices were also crushed. In reality, "civilization" and "Christianity" are so interpenetrated and entangled in Western history that there is no way to cleanly separate them as governing concepts. Each stands for higher values, lofty precepts, refinement, and purity in relation to "the world" of coarseness, brutality, false understanding, and corruption. In the case of civilization—the refined ambience of urban life—its opposite is peasant life and culture. In the case of Christianity—the city of godly believers—its opposite is paganism in its most basic sense. This is not the so-called "paganism" of Greco-Roman civilization, however, for pagan shares a common Latin root, *pagus,* with peasant, having to do with country district or country dweller. The term "pagan civilization" is therefore an oxymoron and will not be used here.

As the ground-clearing edge of Western Civilization in the New World, agriculture was permitted and encouraged to domesticate the wild landscape. These (for the most part) small farms of European immigrants constituted a major American self-image that expanded into a political ideal of Jeffersonian democracy, explicitly linking the capacity for self-government to the self-reliance of small-scale homestead agriculture. In Jefferson's vision, the small farm was the proper human home. But what does the word "farm" mean? Why, of course, it means a quaint and picturesque assemblage of fields, pastures, animals, gardens, buildings, tools, and people that all together produce abundant and nutritious food. So a farm community would then be made of many farms, offering us the traditional cultural patchwork quilt of hard work, honesty, wholesomeness, repose, and long-term cultural stability—lovely small farms nestled snugly and cleanly in nature.

Here's the real definition, as taken from *Webster's*: "*farm* n. [ME. *ferme* rent, lease fr. OF. *ferme* a lease, leased farm fr. *fermer* to make a contract . . . .] 1. A sum or due fixed in amount and payable at fixed intervals by way of rent, tax, or the like. 2. A fixed sum payable at fixed intervals, as yearly, by a person in lieu of taxes or other dues which he has authority to collect . . . also, collectively, the farmers of public revenues." And so it goes until we get to number 6: "A piece of land held under lease for cultivation." In plain English, a farm is a revenue-collecting tract or unit. Let that sink in a moment. For here, beneath our common, everyday understanding of "farm," is the key to the history of agriculture and its present tormented condition within the industrial economy.

Now a *farmer* is: "1. One who farms; as a) One who takes taxes, customs, excise, or other duties . . . as a farmer of the revenues . . . . b) One who leases a government monopoly. c) A lessee. d) A cultivator of ground as a steward or tenant. e) One who conducts or manages a farm. f) One who agrees to perform certain duties for a fixed sum; specif., one who agrees to care for, or to keep, babies or paupers. 2. A clumsy stupid fellow; a dolt." So, a farmer is: a tax collector, a tenant, a hired hand, a dolt.

Let's see specifically what *Webster's* will say about *peasant*. We will slide over the Middle English and Old French and go to its Latin root, *pagus*, whose shade of meaning has to do with the "native religion of country districts and remote places." In other words, the peasant (the dolt) and the pagan (the country bumpkin with questionable beliefs) come from the identical Latin turf, *pagus*, the village or country district. (We could also explore how the use of the word *pagus* drove Christianity out of Creation and into the cathedral, from dusty sandals to bishop's miter, from parable to creed, from lowliest of servant to doctrinal infallibility, and how technological progress, not content with the fruit of knowledge of good and evil, is already into sustained analysis of the Tree of Life, but this is sail for another wind.)

Now, of course, to be *civilized* is a good thing. We all are. We despise, or maybe merely look down upon, those who aren't. To be *civil*, to be *civilized*, to stand for *civilization*, is to have had our sensibility, our very ways of perceiving reality, shaped by the high culture prerogatives of the *city*. All these have a common root. And what is *civilization*? (Here's where our eyes begin to glaze over, for one of two primary reasons. Either such thinking is too hard for our ahistorical, media-saturated brains, or we begin to experience a kind of moral vertigo and intellectual panic that must be avoided, for reasons of identity, at all costs.) Civilization is a tremendously complicated assemblage of urban institutions and governing structures that came into existence primarily on the abundance of agricultural production in the Late Neolithic.

No, the Late Neolithic is neither standup comedy nor a midnight talk show. It is that fabulous period, from six to ten thousand years ago, when (the anthropologists tell us) women turned gathering into horticulture, and men thrust hunting through horticulture to domesticate cows, sheep, goats, pigs, and horses. Thus was common farming developed. Horticulture triggered abundance in a powerful new formation, and it wasn't long before the villages, densely populated and richly provisioned, became the prey of bandits. Aristocracy means, literally, the rule of the strong. How did the "aristocracy" come into possession of the land? How did the peasants become serfs and slaves whose production was stolen? By pillage and by seizure, by force and by sword point. (Tax, from the Latin *taxare*, to touch sharply.) The point is, civilization with its kings, princes, dukes, and armies also had its farmers—that is, its tax collectors, who expropriated the production of the peasants. Civilization, mind you, has *always* existed precisely in this way; with the threat or practice of violence, it has collected revenue, in cash, in kind, and in bodies.

What's different for us is that we're over two hundred years into the industrial revolution, which has been steadily wiping out agrarian culture according to its prevailing technological capabilities and "economies of scale." We are also under the illusion that, as a "democracy," we decide our fate and determine our future by rational discourse and cultural value. We somehow imagine (if we think about it at all) that we have "democratized" the predatory class structure inherent in civilization. And, as an economic system—no contradiction here if we simply ignore it—we pride ourselves in "letting the Market dictate true value," and in "waiting for Science to teach us the future." What *is* the value of an ore body deep in the ground? Of a barrel of oil? A gallon of gas? A forest? A tree? What's a bushel of wheat worth? Or a gallon of milk? Aha! The Market will tell us, without bothering to complicate the secret invisible-hand formula with factors for pollution, depletion, cultural dislocation, environmental abuse, or the consequences of long-term global warming. Not to speak of a fair return, parity, for the independent small producer.

As bigger machines, driven by internal combustion engines, have come into existence, and as farmers have purchased these machines, fewer farms are required because one agribusinessman can do the work (achieve the production) of ten or fifty or one hundred or five hundred farmers. Add fertilizers and genetic engineering and NAFTA and NAFTA cloning. The WTO. And, with supply rationalized to the latest optimal maximum, the monopoly infrastructure producers (tractors, fertilizers, et al.) can charge "what the Market can bear" and the monopoly processors (grain companies and milk, meat, and vegetable processors) can squeeze down the price paid for "raw materials" to the lowest possible margin. Add in some campaign contributions to key Senators and Representatives.

We have neglected one word in particular. That word is *culture*. Now, in *Webster's* definition of civilization, there is a distinction made in regard to culture. Allow me to quote it: "CIVILIZATION applies to human society, and designates an advanced state of material and social well-being. CULTURE, as applied to society, emphasizes the intellectual aspect of civilization; as applied to individuals, it suggests such enlightenment as is acquired by intercourse with what is best in civilized life, esp. as this enlightenment evinces itself in delicacy of taste and nicety of breeding. The term is occasionally abused by a somewhat arrogant and exclusive restriction of its meaning." The crimes of civilization, its very foundation of continual expropriation and denial of expropriation, hide behind this haughty contempt, a presumed superiority in every worthwhile aspect of living. Those who have been the recipients of expropriated largesse have claimed Culture as the preserve of their tastes and manners. They have gone so far as to assert, without apparent irony, that those who fall outside of and below their restricted circle of class do not even know the meaning of culture, much less have any. That's the sort of sustained civilized criminality that has given *culture* a bad name in the countryside, and that has robbed the farming community of a terribly important word. The Latin is *cultura,* from *colere,* to till or cultivate. "1. Art or practice of cultivating; manner or method of cultivating; tillage. 2. Act of developing by education, discipline, etc., the training or refining of the moral and intellectual nature." When Joel Dyer, the chronicler of the modern farm depression and its militia desperation, quotes Val Farmer, a rural sociologist—"To lose a farm is to lose part of one's own identity. There is probably no other occupation that has the potential for defining one's 'self' so completely"[1]—he is talking about *culture,* the profound historic embeddedness of farming as a way of life. The overwhelmingly neglected question in the "farm crisis" is precisely this: what is the historic *cultural* value of small-farm communities and agrarian life?

Have we ever had a real democratic discussion in this country about the cultural and social meaning—the price!—of losing, for the first time in recorded history, small-farm culture and peasant life? No, of course not, at least not since the People's Party of the late nineteenth century. We don't go in for that sort of wussy intellectualizing or irrelevant historical gobbledygook. We believe in dollar bills, unleaded gas, jet skis, and professional football, thank you.[I] The entire concept of culture has been so cornered by upper class snobbery and civilized presumption that the very word on which to hang the defense of rural life has been lost to us. And there is no other word, no other comparable concept, that contains the necessary complexity of meaning. So, lacking an agrarian tradition with real intellectual and political teeth in it, and lacking class allies (labor unions are for cheap food, for cheap food stretches the weekly paycheck) farm communities bite the dust. So we live on the cheap: cheap raw materials generally, including gas and oil; cheap sweatshop imports of clothing, shoes, and electronics; while the upper reaches of the economy are awash in cash, which has inflated the stock market over one thousand percent since 1980! What's left of *thousands of years* of peasant culture is dying in our midst, in our very own time. (By comparison, the rise and fall of so-called communism is of minor historic significance.) We enter the twenty-first century with a landscape terribly disfigured by wildly inflated expectations of our mastery over nature and by untended barns falling in on themselves.

There are really two outstanding and intellectually unanswered questions to ask. The first is, doesn't agrarian culture deserve to live and shouldn't we, democratically, see that it does live, with an adequate population and an adequate income? The second is, whatever makes us believe that there won't be massive chaos and even starvation *in this country* when the inevitable economic, ecological, and/or political catastrophes hit and there is no small-farm culture to cushion the blow, no *real* safety

I. But for a deep and sympathetic look into agrarian troubles in the last quarter of the nineteenth century, see Lawrence Goodwyn's *The Populist Moment.*

net? Or are we God's chosen people operating in God's chosen system in the utopian futures market of technological perfection, in *control*?

NOTES

1. Dyer, *Harvest*, 34.

11

A Sea Change of Red-Green Babies

Green politics, like the African National Congress (ANC) in the decades between World War II and the collapse of the Soviet Union, faces an intransigent opposition, though there are many ways the comparison breaks down. Greens are not an oppressed and disenfranchised majority, struggling for equal rights. Greens are not being threatened, beaten, banished, jailed, or killed simply for calling attention to their life condition and political agenda. Greens are not denied good housing, advanced education, or a decent job just because they're Green. They are not prevented from having sex with, or marrying, Reds. There are no repressive laws against political love affairs across ideological lines, no laws preventing the mating of stewardship and sharing. One could conclude—many people do—that Greens are pampered suburban kids whining about the excesses of their advantages, exactly the opposite of black South Africans in the formative years of the ANC. Or, more cruelly, Greens are snotty, yuppie, spoiled liberals lacking either a real work ethic or patriotic guts.

While it's always possible to find individuals who roughly match the derogatory stereotype in any political song and dance, Greens, in their deepest grasp of political ecology, are an incredibly perceptive and brave lot. They are far less like suburban kids complaining about the noise of the Chemlawn truck than they are subsistence fisherpeople in a rowboat trying to deflect an oblivious aircraft carrier from dumping excess fuel and the ship's slops into a delicate fishery. Greens are asserting what has become increasingly obvious: human beings, the "highly civilized" kind above all, are fouling the ecological nest on a global scale. But this stance, while clearly true and becoming ever more clearly true, bobs in the waves as a relentlessly heedless oncoming aircraft carrier of incredible historical momentum threatens to utterly crush the feeble protest. The prospects do not look bright for the Green fisherfolk in the rowboat.

Saving metaphors are at hand—a turning of the tide, a sea change of consciousness—but this is not a novel or a movie with a prescripted Happy Ending. I write this on Tuesday, March 18, 2003. Last night, George W. Bush, reviewing the "threat" Iraq poses to the West, especially the U.S., gave Saddam Hussein and his sons forty-eight hours to get out of town, leave Iraq, or the Bush posse will roll in and tear the place apart. So we lurch, with magnificent stupidity, toward World War III.

Politically speaking, there are not one, but two, rowboats with fisherfolk bobbing in the waves over the delicate fishery, as the aircraft carrier bears down. One rowboat has Greens in it. The other has Reds. The Greens are saying "Don't pollute this fishery!" The Reds are saying "Because we all like to eat fish!" The mating of Green and Red is now what it's all about, the natural breeding of Red-Green babies the world over, in every clime and culture. Red says "We've got to share." Green says "Only within ecological limits." This is the view from the rowboat, the view from the garden. In our heart of hearts we know this is true, true, true—without the slightest shred of doubt. Bush's posse knows it's true. The aircraft carrier captain knows it's true. Everybody knows it's true. But, since we have a political religion that insists that sharing is evil, that a political program of sharing is inherently wicked, we vote for selfishness and call it Freedom. We cheer on war and call it Peace.

If you want on the winning side, either of the rowboats has seating capacity and is the place to be. Red-Green kids are emerging as the wave of the future. A sea change of consciousness is about to occur. The tide's about to turn. History may not be prescripted, but its unfolding is, to some extent, predictable—at least insofar as there are people with adequate spiritual groundedness, like Lewis Mumford and E. F. Schumacher, to divine the underlying dynamics. Perhaps even Isaac Asimov.

12

Isaac Asimov's *Foundation* Trilogy

An Unanticipated Future

As literature, science fiction can be pretty thin gruel, its characters one-dimensional and shallow, its themes bizarre, and its plots, one might safely say, completely out of this world. Science fiction seems a species onto itself (the middle ground may be utopian novels like *Looking Backward* or dystopian novels like *1984*), and perhaps it is inappropriate to even compare it to traditional literature; even the best of science fiction can't hold a candle to Eliot's *Middlemarch*, Hardy's *Tess*, or Lawrence's *Sons and Lovers*. Yet science fiction has a peculiar vitality all its own, and it would not only be snobbish but blind to ignore the insights of science fiction writers.

Science fiction has been amazingly accurate, even uncanny, in its predictions of future phenomena—what Lewis Mumford in "Progress as 'Science Fiction,'" Chapter 8 of his *Pentagon of Power*, calls "an almost clairvoyant fore-knowledge of present-day society."[1] That that has been the case in technological terms is clearly true. One need only remember the Flash Gordon comic strips with their rocket ships and travel to outer space to get the point. Mumford, referring explicitly to Bellamy's *Looking Backward* and H.G. Wells' *Modern Utopia*, says that although these writers

> . . . possessed an immense amount of useful knowledge about the physical constitution of the universe and the fabrication of machines and machinelike organizations, they showed almost no apprehensive insight into the repeated miscarriage of human purposes that has resulted from the early practice of reducing men to the status of machines. To build up human autonomy, to control

> quantitative expansion, to encourage creativity, and above all, to overcome and finally eliminate the original traumas that accompanied the rise of civilization—of these fundamental needs there is no utopian hint.[2]

Though I have read relatively little science fiction—some each of Vonnegut, Heinlein, Clarke, LeGuin, and so forth—the plot device of Isaac Asimov's *Foundation* trilogy has come forcefully back to me after years of dormancy. It is as if Asimov's mind settled, perhaps unconsciously, on a basic dialectic of human experience and turned it into an instrument of prophecy. Or so, at least, it seems to me. This basic dialectic could be expressed as a centralist drive versus a decentralist impulse, or as an urban consolidation as opposed to an agrarian dispersal. It could also be compulsory grouping as against voluntary association. Yet none of these dialectics is entirely adequate.

In Asimov's long and involved story, the essential structure is simple and uncomplicated. The first book opens on the planet Trantor, seat of the Imperial Government of the Galactic Empire. As Asimov says, "All the land surface of Trantor, 75,000,000 square miles in extent, [is] a single city." Its population, at its height, was "in excess of forty billions."[3] This Galactic Empire, expanding and politically coherent for thirteen millennia (one can see that Asimov is either writing in a typically big way, common to science fiction, or he may be roughly alluding to the duration of human civilization on our very own Earth), has reached the point of rapid decline. Yet this decline is hidden from or ignored by the vast masses of people, including the political leaders. These very leaders, in fact, consider it treasonous when one man, the psychohistorian genius Hari Seldon, predicts the imminent deterioration and breakdown of the Galactic Empire. Asimov's trilogy—or at least its first volume—was copyrighted in 1951. The conception of a "psychohistorian" may then have seemed farfetched, as it, in fact, still does. Yet within ten years of Asimov's first volume, Norman O. Brown would publish his classic and scholarly *Life Against Death*, with its amazing subtitle—"The Psychoanalytical Meaning of History." Asimov's Hari Seldon was at root a mathematician; his calculations of "psychohistory" were based on "the reactions of human conglomerates to fixed social and economic stimuli."[4] Seldon's psychohistory was a complex statistical projection of trends inherent in mass behavior, whereas Brown's psychoanalytical interpretation was based on Freudian symbolism with its explosively rich poetic imagery. The differences between the two are distinctive, to be sure, but the similarity of terminology is startling.

But back to Trantor, where the only greenery on the entire planet is the "one hundred square miles of natural soil, green with trees, rainbowed with flowers" set around the Emperor's palace—"a small island amid an ocean of steel . . . ."[5] The Galactic Empire, in Hari Seldon's analysis (which is, of course, the correct analysis), is operating out of mere habit and accrued inertia. As the gifted psychohistorian points out to his antagonists within the Empire:

> The fall of Empire . . . is a massive thing, however, and not easily fought. It is dictated by a rising bureaucracy, a receding initiative, a freezing of caste, a damning of curiosity—a hundred other factors. It has been going on, as I have said, for centuries, and it is too majestic and massive a movement to stop.[6]

Seldon accordingly predicts a period of rapid decline and eventual collapse, and he sets about establishing two "foundations," as he calls them, as a means of maintaining a vital civilizing impulse. In Seldon's calculations, collapse of Empire without corrective intervention would result in 30,000 years of barbarism. With intervention—namely his own through the instrumentality of the Foundations—Seldon sees that the period of barbarism may be shortened to a mere 1,000 years. At that point a new Empire, based on the unified Foundations, should arise. Seldon maneuvers the hostile Empire leadership into exiling his First Foundation of twenty thousand families to Terminus, an uninhabited planet lacking metals, at the edge of the Galaxy. The Second Foundation, of which only his closest associates are aware, Seldon will only vaguely allude to; he says it is to be established at the opposite end of the Galaxy from Terminus, at a place called Star's End. Most of the first book, and the bulk of the second, are given over to the struggle for political viability and then expansion of the First Foundation.

The alleged purpose of the First Foundation—that of collecting and recording all human knowledge in one great Encyclopedia Galactica—is revealed after fifty years to be a hoax: Seldon's real purpose was the establishment of a dynamic political entity that could expand through the conquering of barbarous planets even as the influence and control of the Empire declined. This purpose it carries out according to Hari Seldon's calculations. It is instructive to note, however, that only one psychologist was among the original exiles to Terminus. The purpose of the First Foundation was political and technological; it was to act, not out of deep psychological insight, but by the guiding impulse of political survival. It was, in other words, an instrument of statistical probabilities. The crises the First Foundation faces are, naturally enough, referred to as "Seldon Crises," and the leaders understand enough of Seldon's purpose to grasp that such crises "are not solved by individuals but by historic forces." Eventually one leader, Hober Mallow, realizes that "The whole war is a battle between . . . two systems; between the Empire and the Foundation; between the big and the little."[7]

In the end—the end of the trilogy, that is, not the end of the thousand years—important members of the First Foundation, having seen first the remnants of the old Empire and then other unexpected challenges swept aside, come to the peculiar, but predictable, conclusion that the hidden Second Foundation is the real enemy. The Second Foundation, in turn, with its vastly superior psychological skills, resolves this threat with its usual alacrity, and the novel concludes on a hopeful note: the First Foundation, having been led to believe, quite falsely, that it has succeeded in destroying the Second Foundation, proceeds on its probable way, again within the calculations of Seldon's psychohistory. And the Second Foundation remains hidden and anonymous, the true inheritor of Hari Seldon's genius.

II

At the beginning of these remarks I said that Asimov's dialectical tension between the Empire and the Foundation, as well as that tension between the separate Foundations themselves, seemed prophetic. Asimov's conception of the Galactic Empire, it seems to me, resembles in some respects the current condition within industrial civilization. Despite the fact that no resource crisis is attributed to the Empire—it has, after all, the entire Galaxy at its disposal—it is instructive to note that the First Foundation is exiled to Terminus, a name suggestive in its own right, and a planet without metals. Hence, the First Foundationers learn to make do with less; the essential scarcity forces them to modify their technology in the direction of conservation and compactness. The remark I quoted earlier—that "The whole war is a battle between those two systems; between the Empire and the Foundation; between the big and the little"—is a theme of the entire trilogy.

The First Foundation, in its long period of expansion, passes through several familiar phases: psychohistory as bunk religion becomes supplanted by domination through trade; and trade gives way to a growing democratic socialism. This advance and growth of the First Foundation is reminiscent of Alexis de Tocqueville's anticipation of the march of democracy in his famous *Democracy in America*:

> The whole book which is here offered to the public has been written under the impression of a kind of religious dread produced in the author's mind by the contemplation of so irresistible a revolution, which has advanced for centuries in spite of such amazing obstacles, and which is still proceeding in the midst of the ruins it has made.[8]

De Tocqueville goes on to express his conclusion that the social leveling generated by democratic institutions would result in a

> . . . nation . . . less brilliant, less glorious, and perhaps less strong; but the majority of the citizens will enjoy a greater degree of prosperity, and the people will remain quiet, not because it despairs of amelioration, but because it is conscious of the advantages of its condition.[9]

But the vision that foresees mere materialistic satiety at the end of the democratic process does not see far enough.

Yet at the close of the third novel, the First Foundation has taken on some of the attributes of the destroyed Empire—it has become materialistic and big, in other words—as it seeks to destroy the smaller and more spiritual Second Foundation. Asimov keeps returning to this theme. He has one character say that the Second Foundation is "efficient in a sense far different from the efficiency of the old Empire or of the First Foundation . . . . All these have brought mechanical efficiency to their subjects at the cost of more intangible values. [The Second Foundation] brings happiness and sufficiency."[10] This particular idea is uttered in the setting of a peasant village called Rossem, located on a remote planet with arctic-like conditions—reminiscent, perhaps, of peasant villages in Asimov's native Russia. Another character, a member of the cabalistic Second Foundation, says that it is "the intention of the [Seldon] Plan to establish a human civilization based on an orientation entirely different from anything that ever before existed."[11] The First foundation is to supply "the physical framework of a single political unit," while the Second Foundation is to provide "the mental framework of a ready-made ruling class."[12]

This latter quotation, regarding a "ready-made ruling class," is, not at all incidentally, an example of the intellectual inconsistency and spiritual immaturity of science fiction. On one page Asimov's character alludes to a civilization "based on an orientation entirely different from anything that ever before existed," and on the next page the same character can all but destroy that fragile conception by the altogether ominous idea of "a ready-made ruling class." All along, Asimov has essentially postulated, implicitly to be sure, something akin to the Marxian idea of the withered state, a kind of society where class domination has ceased to exist, a steady rise of the entire population to new levels of consciousness and grace. So his unfortunate, and I think careless, insertion of "ruling class" terminology really injures his essential vision. (Perhaps this "ruling class" language is not so much a matter of intellectual carelessness as it is the fogginess of spiritual limitation. Asimov—I believe—intuited something, a richer, deeper configuration of human awareness, culture, and governance, that he was intellectually incapable of portraying fully or spelling out. Therefore, he reverted—carelessly? impatiently?—to a conventional concept that betrayed his intuition. Perhaps this is where Buddhist meditation is needed. These are the kinds of spiritual eggs we need to patiently hatch before civilization and its various forms of toxic blowback thoroughly cook our ecological and cultural geese.)

Be that as it may. It is instructive to note that the First Speaker of the Second Foundation is a *farmer*, and an agent for an agricultural cooperative. And the planet on which the Second Foundation is centered is none other than Trantor, the original capital of the Galactic Empire. After the final collapse of the Empire, the citizens of Trantor tore down most of the supercity that had, except for the one hundred square miles of park surrounding the Emperor's mansion, been entirely given over to concrete, plastic, and steel. As Asimov informs us:

> The survivors tore up the metal plating and sold it to other planets for seed and cattle. The soil was uncovered once more and the planet returned to its beginnings. In the spreading areas of primitive agriculture, it forgot its intricate and colossal past.
>
> Or would have but for the still mighty shards that heaped their massive ruins toward the sky in bitter and dignified silence.[13]

The First Foundation is urban and technological; the Second Foundation is agrarian and psychological. This dialectical tension between the aggressive city and the homey countryside has been a consistent and major theme in Western literature, from the Roman poet Virgil to the Kentucky farmer Wendell Berry. One girl who falls under the care and protection of the Second Foundation is even called Arcadia, named after the ancient pastoral district of the central Peloponnesus in Greece, a symbol of rural peace and rustic simplicity.

III

Pastoral imagery tends to emerge in those times and instances where agrarian coherence is disintegrating under the impress of urban or industrial forces. Such was the case in Virgil's time, in the first century B. C. It was true in Wordsworth's time, and it is certainly true in ours—except that in the modern period this agrarian disintegration is *fait accompli*. One of our major intellectual problems is to *rediscover* the substance within pastoral imagery and quit treating that imagery as mere literary amusement. For industrial civilization has pretty thoroughly destroyed agrarian culture, and the problems generated by that destruction are real, not mythical. As Roger Burbach and Patricia Flynn point out in their *Agribusiness in the Americas*:

> Besides *linking* agriculture to industry, agribusiness also means that agricultural production increasingly *resembles* industrial production, in the application of technology to control nature and increase productivity and in the use of wage labor.[14]

From ninety-five percent of the total in 1790, the American farm population is now down to less than two percent.

If Asimov's fanciful depiction of events is essentially prophetic for our own circumstance—and, as I have already indicated, I believe it is—then the appropriate technology and back-to-the-land movements in many Western countries have been important indicators of things to come. It seems possible to interpret Asimov's trilogy like this: the old industrial civilization has outgrown its ecological viability; its cosmic function (let's not get too mystical) is essentially fulfilled; and its collapse is inevitable unless preventative measures are implemented quickly—measures that would not serve to maintain the status quo but make transformation to an ecological order less brutal. At the same time, the industrial and political leaders denounce such ideas as scurrilous, if not worse. The military-industrial-media-political elite are wedded to conceptions of economic growth, commercial protection, and technological expansion. That these conceptions are often contradictory to, if not destructive of, democratic principles and global ecology seems to be rather systematically ignored. Our situation so far seems to parallel Asimov's Empire.

Our world, however, is not a single political entity. We could stretch our minds a bit and identify the former Eastern bloc nations, with their large collective economies, as similar to the mature First Foundation, given the important distinction that all the Eastern bloc countries, with the possible exception of Yugoslavia, lacked any sort of meaningful democratic structure. The ideal of the West, on the other hand, of small-scale enterprises, could possibly be correlated to the inclinations of the Second Foundation, although as every one knows (or should know), small-scale enterprise is now largely a political myth reserved for the mawkish rhetoric of election-year politicians. On another level, both the Empire and, to a lesser extent, the First Foundation as well, are oriented toward an urban-based technology and massive industrial productivity. All modern industrialized nations share these proclivities. The First Foundation, as a form of democratic socialism centered in an urban and (somewhat modified) industrial context, is a construct for what could develop within our own civilizations, both East and West, after the contraction of the present expansive, high-energy system. The *Second* Foundation may seem at first glance to have no parallel in and no relevance to our emerging order. Yet it does, if one looks more carefully and more closely.

Agribusiness is an integral facet of advanced industrial society; economic pressures associated with the industrial system have forced the contraction and virtual elimination of traditional agrarian culture. This is true for both East and West, and is increasingly true—as Roger Burbach and Patricia Flynn point out—in the Third World. As Arnold Hauser has stated so succinctly in his book *The Philosophy of Art History*: " . . . there is no more folk art because there is no longer anything that [can] be called 'folk,' and this is in fact true of the West and especially of the Anglo-Saxon lands, for in them not only the masses in the industrial towns, but also those employed in agriculture have now hardly anything in common with those who kept the former folk art alive."[15]

The breakdown of energy-intensive civilization will require, as in the aftermath of the fall of Empire on Trantor, a reorientation toward a fundamentally *biological* agriculture with a far higher rural population. This is precisely the development that the late E. F. Schumacher vigorously advocated. If rural population could become large enough, it would be able to generate its own distinctive culture; it could become, in a manner of speaking, a genuine second foundation of agrarian autonomy and regional authenticity. Nor should it pass unnoticed that such "psychological" organisms as Mildred Loomis's School of Living, John and Nancy Todds' New Alchemy, William Irwin Thompson's Lindisfarne, Sally and Maynard Kaufmans' School of Homesteading, or the mushrooming of rural communes only a couple of decades ago, were (and, to some extent, still are) already on the scene. They are admittedly insignificant in terms of their size and present constituency; but they are the seeds, one might say, of a new and radically sophisticated agrarian culture.

We can postulate, therefore, social developments that will proceed similarly in our context as in Asimov's trilogy. Our industrial civilization, oblivious of its excesses and unable (or unwilling) to discipline its appetites, resembles in some important ways Asimov's Galactic Empire. The deterioration of our urban-industrial system—whether through ecological dysfunction, resource depletion, economic dislocation, war, or cocktail combinations of these four—will only heighten the already urgent problems of meaningful work and social anomie. The restoration or recreation of agrarian culture seems to me a likely outcome of these conditions. On the other hand, the transportation, communication, and industrial infrastructures cannot deteriorate beyond a certain point before something akin to barbarism sets in. I suggest that within the lifetimes of our children and grandchildren a democratic, ecological socialism will emerge in this country in which public ownership and management of vital services and industries—including *decentralized* public ownership—becomes commonplace.

IV

We are living in a time of enlarging and deepening crises, and our ability to act effectively toward humane and ecological resolutions of those crises depends heavily on the clarity and depth of our analysis. Not all proposals for resolving the malaise of industrial civilization are equally viable; some are not viable at all; and some would only deepen and extend the dilemmas. The string of cataclysmic changes the world has suffered through in recent centuries, especially stemming from the industrial revolution, with its factory mode of production, has not yet reached its end. We are just now, I believe, entering into a final and decisive period of disruption and reorientation of industrial civilization.

In her book *The Death of Nature*, Carolyn Merchant points out that

> Basic to the agrarian ecosystem of premodern Europe was the relationship between the peasant community and the land. Evolved over centuries of adaptation to the productive capabilities of the natural environment on the one hand and the state of agricultural technology on the other, the peasant community produced a level of subsistence by following traditional patterns of cooperation upheld by powerful cultural norms.

She goes on to say that the "transition from peasant control of natural resources for the purpose of subsistence to capitalist control for the purpose of profit" runs parallel to the "transition from the organism to the machine as the dominant metaphor binding together the cosmos, society, and the self into a single cultural reality . . . ."[16]

Leo Marx, in his book *The Machine in the Garden*, an analysis of technology and the pastoral ideal, shows that a good many American writers—explicitly Mark Twain, Henry Thoreau, Nathaniel Hawthorne, and Herman Melville—depict the machine as "a sudden, shocking intruder upon a fantasy of idyllic satisfaction. It invariably is associated with crude, masculine aggressiveness in contrast with the tender, feminine, and submissive attitudes traditionally attached to the landscape."[17] In his *Interpretations and Forecasts*, Lewis Mumford carries this line of thought even further when he says that the historian Henry Adams

> . . . saw that the males of his society, who had transferred so many of their once autonomous activities to machines and automatons, did not have sufficient life-sense to save the race. In their blind pride over their scientific facilities, they would cling to the insensate mechanisms they had created, making them go ever faster and faster, though incapable of applying the brakes, changing the direction, or choosing the destination.[18]

And Jeremy Rifkin and Ted Howard, in *The Emerging Order*, present us with a picture of our society that is at once political and psychological:

> Progress has, indeed, become an autonomous force, rushing inextricably forward with neither a specific destination nor timetable. Human beings, at first willing passengers on the rails of progress, were excited by the great potential adventures that lay

> ahead; now they have become increasingly disillusioned with the journey and their experience. Afraid to jump off for fear of being abandoned to an uncertain fate, and equally afraid to continue on a journey they are convinced is moving toward some terrible crash, they remain frozen and immobilized, unable to act, preferring instead to place their destiny in the hands of forces beyond their reach and control.[19]

In all of these impressive images, one cannot help being struck with the recurring theme of breakdown, catastrophe, and collapse. Industrial society is characterized by collective human behavior shaped and polarized by massive and impersonal economic and political forces. This commercial society has all but obliterated preindustrial, decentralized, and largely small-scale, organic society. We could consider what happened to American Indian cultures in the face of European expansion on this continent. But the issue there is rather obscured by racial and religious intolerances, as well as by ideas of primitive society versus advanced forms of civilization.

For illustrations of the impact of industrialization on agrarian society we should look to the history of Britain, for England, after all, was the first industrial nation in the world. In his *Industry and Empire*, E. J. Hobsbawm says that the industrial revolution destroyed the "traditional world and way of life" of the laboring poor "without automatically substituting anything else. It is this disruption that is at the heart of the question about the social effects of industrialization."[20] Hobsbawm then goes on to describe how

> Labour in an industrial society is . . . different from pre-industrial work. First, it is overwhelmingly the labour of 'proletarians', who have no source of income worth mentioning except a cash wage which they receive for their work. . . . Second, industrial labour—and especially mechanized factory labour—imposes a regularity, routine and monotony quite unlike pre-industrial rhythms of work, which depend on the variation of the seasons or the weather, the multiplicity of tasks in occupations unaffected by the rational division of labour, the vagaries of other human beings or animals, or even a man's own desire to play instead of working . . . . Third, labour in the industrial age increasingly took place in the unprecedented environment of the big city. . . . Fourthly, pre-industrial experience, tradition, wisdom and morality provide no adequate guide for the kind of behavior which a capitalist economy required.

Regarding his third point on industrial consolidation within an urban context, Hobsbawm says bluntly that "the city destroyed society."[21]

We who live in industrial society, where all our economic umbilicals lead directly into the incredibly complex, interconnected and energy-consuming industrial-financial system, usually fail to comprehend the tone and quality of life in preindustrial modes of living. We all too readily accept that famous epithet by Thomas Hobbes—and recently echoed by C. P. Snow—that life before the modern industrial period was always "poor, nasty, brutish and short."[22] We therefore also fail to anticipate the recreation of modes of life outside of or beyond the standardization of the modern system. We seem pretty oblivious to the fact that, as Hobsbawm puts it, "No change in human life since the invention of agriculture, metallurgy and towns in the New Stone Age has been so profound as the coming of industrialization."[23] It is nevertheless true that the annihilation of organic society, with its rich culture of self-provisioning and subsistence adaptability, presents a very grave danger to human survival. The modern state, with its awesome power of destruction and control, has reached such a pinnacle of autonomy precisely because it has destroyed the multitude of decentralized, small-scale human communities with their attendant cultures; the modern state has broken low-energy, organic society to pieces and grown to mammoth proportions on the individualized, routinized, and urbanized remains. We can anticipate the creation of new forms of organic society only on the basis that industrial civilization will suffer some sort of breakdown or decay. Only the serious deterioration of the present industrial system can create the functional conditions and the psychological freedom in which a new culture might emerge. The hegemony of the industrial city must be undone.

Asimov's conception of two distinct "foundations" emerging from the breakdown of technocratic civilization is profoundly insightful. We have, on the one hand, a spokesman for the "first foundation" in the person of the democratic socialist Michael Harrington, who says that the future is going to be collective; it's only a question of whether that collectivity will be democratic and humane or severe and authoritarian. On the other hand, we have a woman like Mildred Loomis of The School of Living who points the way to the creation of new patterns of society outside the aegis of urban political collectivity. We have had, roughly, the democratic socialists on the one side and the so-called counterculture on the other, with all its New Alchemys, Lindisfarnes, Michigan Land Trusts, Schools of Homesteading, and (ancient history) *Whole Earth Catalogue*.

I've had a foot in each camp, psychologically speaking. I believe both "foundations" are vital to our cultural integrity, yet my concern at this point is primarily directed toward the "second foundation." We are all so psychologically and emotionally neutralized, so culturally bewildered, by the power, productivity, and sheer entertainment capacity of the political-industrial-media utopian circus that we must constantly test and challenge our passivity and comfortableness. We no longer know the meaning of "society" except as an abstraction; and when the word "community" can be used to refer, for example, to the "military-industrial community," we are essentially acknowledging that we no longer have any conception of what community is about. A crisis is building in the interface between global ecology and economic globalization, and people become reactionary and even fascist in the face of breakdown, primarily because it is the path of least resistance. Casting about for scapegoats may be possible until the day of true reckoning arrives, but then all the carefully avoided decisions will face us en masse, multiplied in intensity because so long unattended to. We cannot have a living, vibrant, earthy culture by consenting in an ongoing way to be merely the system's personnel—or we will end up, as Rifkin and Howard point out, "frozen and immobilized, unable to act, preferring instead to place [our] destiny in the hands of forces beyond [our] reach and control."

It is interesting, in one truly major respect, to see how miserably Asimov's Hari Seldon fails to anticipate the future. In the entire trilogy, women play very few substantial roles. Yet here we are, barely half a century past the publication of Asimov's work, and the women's movement has already assumed a truly remarkable influence on society. Nor is this movement merely incidental to the crisis of industrial civilization. As Carolyn Merchant has pointed out:

> The ancient identity of nature as a nurturing mother links women's history with the history of the environment and ecological change. The female earth was central to the organic cosmology that was undermined by the Scientific Revolution and the rise of a market-oriented culture in early modern Europe. The ecology movement has reawakened interest in the values and concepts associated historically with the premodern organic world. The ecological model and its associated ethics make possible a fresh and critical

> interpretation of the rise of modern science in the crucial period when our cosmos ceased to be viewed as an organism and became instead a machine.[24]

Theologian Rosemary Radford Ruether, in her *New Woman/New Earth*, has shown that "Sexual symbolism is foundational to the perception of order and relationship that has been built up in cultures." This insight allows her to insist that we are now in a period "when the women's movement, properly understood, encompasses all other liberation movements."[25] And although Barbara Epstein (in an essay entitled "What Happened to the Women's Movement?") laments both the lack of a "clear political agenda" among contemporary feminists and the substitution of "aspirations toward material success for the demand for social equality and community,"[26] the slow but steady rise of women into positions of public power and policy-making *will* inevitably result in a society in which feminine values determine public policy.

The freeing of "feminine" nature from the exploitative grip of the "masculine" industrial machine implies not only the full release of women's energy into the public sphere, it also implies the restoration of organicism as the overarching metaphor for society as a whole. A mathematically-based psychohistory might never have anticipated this development, nor have been able to explain its symbolic relevance. Yet psychoanalysis, based on new understandings of the depth and complexity of the human mind and shed of its sexually prejudicial features, has done both. Hari Seldon should have known that mass behavior is strongly influenced by a symbolism that links human consciousness, and even human happiness, with the well-being of nature.

A small flaw, perhaps. The "second foundation" will surely have to take this new factor into account. The stone that the builders rejected, said a Semitic sage, becomes the corner of a new society. A small oversight—but one that might yet save us all from nuclear extermination. Meanwhile, the restoration of organicism as our overarching metaphor requires a greening of the economy, and a greening of the economy requires a major reconfiguration of the elements within economic theory. A Green revolution in outlook and behavior both implies and requires a massive transformation in how we obtain our food, clothing, and shelter.

NOTES

1. Mumford, *Pentagon*, 212.
2. Mumford, *Pentagon*, 219.
3. Asimov, *Foundation*, 12.
4. Asimov, *Foundation*, 17.
5. Asimov, *Foundation*, 15.
6. Asimov, *Foundation*, 29.
7. Asimov, *Foundation*, 198.
8. Tocqueville, *Democracy* (Vol. 1), 7.
9. Tocqueville, *Democracy* (Vol. 1), 10.
10. Asimov, *Second*, 45.
11. Asimov, *Second*, 91.
12. Asimov, *Second*, 92.
13. Asimov, *Second*, 155.
14. Burback and Flynn, *Agribusiness*, 12.
15. Hauser, *Philosophy*, 330.
16. Merchant, *Death*, 43–44.
17. Marx, *Machine*, 29.
18. Mumford, *Interpretations*, 363.
19. Rifkin and Howard, *Emerging*, 94.
20. Hobsbawm, *Industry*, 84.
21. Hobsbawm, *Industry*, 85–87.
22. Snow, *Two*, 44.
23. Hobsbawm, *Industry*, 21–22.
24. Merchant, *Death*, xx.
25. Ruether, *New*, 3, xi.
26. Epstein, "What," 10–11.

13

Green Thoughts on Economic Theory

In modern Economic Theory, Land, Labor, and Capital have been designated the primary elements of production. Land, in this designation, includes the whole of the natural world; it takes in all the forms and residues of life; it encompasses Air, Water, Fire—the three elements that, along with Earth, constitute the fundamental ingredients of life in the thought of ancient cultures. Labor, in the modern framework, does not refer to people generally or to reproductive labor specifically. It means deliberate production by paid personnel for the market and mass consumption. As a category within Economic Theory, Labor is only as old as industrial civilization, out of which the "science" of economics arose. Labor, in this respect, is linked closely with an exchange or circulation of money; that is, Labor is purchased human energy in an economic context in which virtually all necessities, luxuries, and armaments are forced through the monetized market prior to consumption.

When Labor is replaced by machinery, automation and robotics, the new form of productive energy becomes Capital. Modern Capital, through innovative technology, makes Labor obsolescent and Land obsolete. As Aldo Leopold said in *A Sand County Almanac*: "Your true modern is separated from the land by many middlemen, and by innumerable physical gadgets. He has no vital relation to it; to him it is the space between cities on which crops grow."[1] Or, as Harry Braverman put it in *Labor and Monopoly Capital*: "The more science is incorporated into the labor process, the less the worker understands of the process; the more sophisticated an intellectual product the machine becomes, the less control and comprehension of the machine the worker has."[2] Now Capital is the totality of those things (buildings, machines, tools, technologies, systems—the infrastructure, in short) and the "medium of exchange" (money in its various forms) that govern actual productivity in any capitalist economy, corporate or state. Since Economic Theory only pertains to the formal or monetized economy (backyard gardens in the United States produce an estimated twenty billion dollars worth of food each year, but this is only a nuisance and distraction to Economic Theory), functional Capital in the form of tools, machines, buildings, trucks, tanks, bombs, and so on, is considered *as* Capital or *owned* by Capital.

Of the two basic forms of Capital—the actual infrastructure on the one hand, money on the other—money is the most elusive, fluid, volatile, and peculiar. Money can "create" jobs. This is the official rationale behind tax cuts for the wealthy: that the owners and managers of Capital will invest this concentrated energy in productive industries and, by so doing, create more jobs and increase the standard of living as the economy expands. Without the energizing directives of Capital, it is asserted, Labor would come to a complete standstill. (Serfs without the manor lord have no motivation . . .) Without Capital, it is assumed, productivity would stop. Without money and the directives of Capital, there would be nothing to do, nothing to watch on television, and nothing to eat. Without Capital, Labor has no motive and Land is without value.

Now, Labor is the "Bodily exertion or effort directed to supplying society with the required material things; the service rendered or part played by the laborer, operative, and artisan in the production of wealth, as distinguished from the service rendered by capitalists or by those whose exertion is primarily and almost entirely mental." *Webster's* linking of capitalists to "those whose exertion is primarily and almost entirely mental" is supported by the etymology of Capital; the word derives from the Latin *caput*, meaning head. (Cullen Murphy, in *Are We Rome?*, says "class stratification of Roman society was extreme . . . . Rome's wealthiest class, the senatorial aristocracy, constituted by one estimate two thousandths of one percent of the population; then came the equestrian class, with perhaps a tenth of a percent. Collectively these people owned almost everything."[3]) Capital, in Economic Theory, is a concept that crystallizes the "ideal planning" of utopia; it is the concentrated and accumulated energy owned and directed by the hierarchical elite. In a natural sense, Land is as old as the Earth itself, as old as life in all its forms and residues. Labor is certainly as old as life, for all life needs to eat, and the eaten generally doesn't volunteer without persuasive exertion on the part of the eater. Life needs to sustain itself, and that invariably involves effort. Life *lives* on life; and all life, one way or another, resists being consumed, resists death, resists extinction. Labor is the energy expended in the effort to live, consume, and reproduce. For the vast bulk of human history, this effort was a highly personal "hands-on" subsistence experience. But utopian society (especially since the rise of modern industry) has devised manipulative technological complexities by which food emerges magically from money. Any child with a quarter knows this to be true. Or, to put it in more analytical terms, Land no longer produces food, nor does Labor. Capital does. This is, in essence, the production and reproduction of things through the medium of the utopian machine; and this machine is Capital.

Capital in the sense of tools dates back, obviously, to early human society. It is probably true that tools, while the products of collective evolutionary intelligence, were essentially considered personal property, but without what we understand as legal sanction. Property could become *legally* private only with the immense hierarchical structure of that recent human creation, civilization. Formal law creates precise boundaries enforceable through legal institutions and armed force. It may be that machines go back in time nearly as far as tools, if we consider levers, mortars and pestles, or bows and arrows as early forms of machines. The designation is ambiguous in its very etymology, for the word derives from the Latin *machina,* meaning device or trick. The word machinate, to contrive a harmful scheme, derives from the same root. But *machina,* the Latin word, comes from the Greek *mechos,* meaning expedient. Mumford, of course, asserts that the expedient machine of many moving parts has its prototype in the *human* machines of the earliest civilizations: the irrigation workers, the troops of the army, the pyramid builders. These expedient machine people (whom we would today call personnel) were probably slaves. Technically, they were Capital, owned and operated, bought and sold.

Money, too, is considered by historians of antiquity to have been in existence far back in human society. Money or its prototypes, some special and magical objects, held a value or was invested with meaning quite beyond its intrinsic natural worth; it was specially endowed with symbolic energy; to have it was to hold a heightened power, to feel more powerful and secure. Money was, one might say, a kind of magic that developed a social function, a power to shape patterns of economic conduct. And as the coins of early civilizations were often stamped with an image of divinity, then money has had—and continues to have—subtle connections to the realm of the sacred and the transcendent. "In God We Trust." But money also carries a hidden sexual symbolism. Cowrie shells, for instance, which resemble the female vulva, were a form of currency in some early cultures. Freud linked money to an infantile fascination with feces. But it is my belief that Freud was wrong—or, if not wrong, then only partially correct. Money is sublimated sexual energy. It may well represent both female fertility and male potency—penis and womb, egg and sperm. But money may also be gender-specific in its utopian or civilized form. (Anecdotal evidence can be found, for instance, in *Confessions of a Stockbroker: A Wall Street Diary* by a writer who calls himself Brutus: stockbrokers, Brutus says, "tend to be hypermale, equating money with power, and with the penis, if you must carry things to their logical conclusion." Brutus also says that men "equate adventures with money and adventures in sex," but women "often use money and the stock market in quite a different way, as a club against their husbands and fathers. Stock market activity for them is always a rebellion *against* sex, a liberation from the sexual roles forced upon them by the male."[4] The unabridged *Webster's* is suggestive in possible connections between the Anglo-Saxon *geld*, meaning tribute or payment, and the Middle English *gelden*, meaning barren, not giving milk, but also castrated and deprived of anything essential. So to "geld" a people or a territory is to tax it of its essentials, its money, milk or sperm, its wealth, fertile women or virile young men.)

But Land, Labor, and Capital as theoretical constructs in Economic Theory grew out of early modern capitalism. Yet, interestingly, the first "complete system" of economics, according to the *Columbia Encyclopedia*, was propounded by eighteenth-century French physiocrats. (The term physiocracy comes from the Greek words *physis* and *krates* that mean, respectively, nature and partisan. A physiocrat was, then, a partisan of nature.) The *Encyclopedia* says that the founders of physiocracy believed

> . . . all wealth originated with the land and that agriculture alone could increase and multiply wealth. Industry and commerce, according to the physiocrats, were basically sterile and could not add to the wealth created by the land. They did not advocate that industry and commerce be neglected in favor of agriculture, but they tried to prove that no economy could be healthy unless agriculture were given the fullest opportunity.[5]

It seems significant that economics as a special intellectual discipline was first a doctrine of Land wealth. And, indeed, Land as the basis of wealth reflects the original meaning of the word economy—from the

Greek *oeconomia,* meaning household "management," people provisioning themselves in household and village from the land and the commons. But physiocracy took shape in a country where the corrupt and oppressive feudal state was bursting at the seams with the misery of the common people. Peasants "formed at least three quarters of the population of the kingdom," says Georges Lefabvre in *The Coming of the French Revolution*: "Of the land around Versailles peasant ownership accounted for no more than one or two per cent. Thirty per cent is a probable average for the kingdom as a whole."[6] Physiocracy was not an intellectual apology for feudal conditions or the manorial system; rather, it postulated a vision of economic conduct that would modernize farming and monetize the economy.

Economic Theory was quickly picked up, expanded, and modified by the English thinkers Smith, Ricardo, and Malthus. Adam Smith talked with the leading physiocratic theorist Francois Quesnay in 1764 or 1765. Once Smith was back in England (Smith and Quesnay had talked in Paris), Smith transformed French formulations on Land wealth in a way that reflected the emerging industrial structure of England. In 1776, Smith published his *Wealth of Nations*. The industrial revolution was getting underway. First Labor and then Capital was elevated into the wealth-producing throne, into the key and most important position. Economic Theory reflected the realities of political power, the shift in the forces of social organization and cultural structure from feudal conditions to mercantilism to entrepreneurial industry to capitalist consolidation.

Each component of Economic Theory has had its corresponding political expression. Land was held to be primary, as we have seen, by the French physiocrats. Land remained primary for the leading American political philosopher, Thomas Jefferson. Jefferson, himself a slave owner, feared that democracy would fail if uprooted from a close relationship with small-scale agriculture. (In part, this was a linking of economic self-reliance with political independence, with the latter dependent on the former. And, indeed, it is possible to correlate the decline of stable small-farm communities to the rise of cultural standardization. Folk culture everywhere has been characterized by village self-provisioning and an essentially subsistence-oriented life that utilized the market or cash economy infrequently, if at all. Civilization, on the other hand, rationalizes all forms of production and expropriates as much of that production as possible. In the industrial revolution, civilization refined its prototype machine and used it to suck folk culture dry.)

Capitalism contends that ownership of the means of production (Land, access to a Labor *market,* industrial infrastructure, the advertising and education industries, and large reserves of fluid money) should be held by private individuals or in privately-owned corporations. These persons or corporations will then enlarge their stock of Capital through profit extraction, acquisition and merger, and will, as an inevitable consequence, generate more jobs—make room for more Labor—through expanded production by reinvestment. Of course, a substantial portion of this wealth will be siphoned off for stylish living by the owners and top managers as deserved reward for their creativity and risk. (Since it takes money to purchase commodities in a capitalist economy, people must be somehow provided with cash, or its equivalents, even if unemployed. But pure capitalism no longer needs the shell-game circulation of money to expropriate wealth when all productive infrastructure is owned by, or has itself become, Capital. The crisis of capitalism is, in one sense, the problem of money. Or, to put it differently, capitalism no longer needs a huge pool of consumers. In fact, when people as Labor are no longer needed, capitalism can suddenly become "environmentally conscious" and begin stressing the problems of overpopulation, although the current regime of so-called Christian conservatives has, as Kevin Phillips puts it in *American Theocracy*, a "preoccupation" with "maximizing the potential soul count for the hereafter,"[7] and is therefore disinclined to advocate birth control or population limitation.) But overall, the prevailing theory holds, wealth will "trickle down" to everyone and the economy will be "healthy."

Yet, if we look at the economic history of the past couple of centuries, we can't help but be struck with the shrinkage of Land in Economic Theory. This shrinkage correlates to the demolition of peasant culture and small-scale farming, with the explosion of industrial Labor under the control of consolidating Capital. The repudiation of rural culture, physiocracy, or Jeffersonian pastoralism has less to do with the inherent intellectual assets or liabilities of those cultural sensibilities than with the raw power of utopian Capital to shape society according to its ideal planning, peddling its ideology through economic priests. In this context, in the words of Harry Braverman:

> Work ceases to be a natural function and becomes an extorted activity, and the antagonism to it expresses itself in a drive for the shortening of hours on the one side, and the popularity of labor-saving devices for the home, which the market hastens to supply,

> on the other. But the atrophy of community and the sharp division from the natural environment leaves a void when it comes to the "free" hours. Thus the filling of the time away from the job also becomes dependent upon the market, which develops to an enormous degree those passive amusements, entertainments, and spectacles that suit the restricted circumstances of the city and are offered as substitutes for life itself.

Thus the "care of humans for each other becomes increasingly institutionalized,"[8] and new service industries are created to fill the cultural void.

Political power shifted from the landed gentry to urban industrialists, and the folk culture of the countryside was smashed in the process. From cottage weavers to field laborers, the old culture came to an end, crushed between the rise of industrial agribusiness and the factory mode of production. The changing focus of official Economic Theory reflected the power transformation of economic organization, a new consolidation of utopian power and intent. From this we might rightly conclude that Economic Theory, like the technology of utopia, is neither "neutral" nor "objective," but mirrors the power structure of legal ownership and political control.

Green economics has yet to adequately contend theoretically with market factors in an ecological society. On the one hand, an ecological society is neither intensely consuming nor intensely polluting. On the other, the global perspective of Green politics promotes a great deal of cultural exchange worldwide. An ecological society cannot afford agribusiness or a "cheap food" policy or resource plunder of any sort. This implies recycling, durable goods, mass transit, renewable energy sources, and a great deal of localized gardening and small-scale farming. Rural population would rise dramatically in a Green society and much of its production in food and craft would never enter the market nexus. This kind of economic activity is at least one major factor in subsistence-oriented folk cultures. Green economics is, then, a radical eutopian departure from the utopian Economic Theory of either communism or capitalism. Contemporary Economic Theory has prided itself on its stand of "objectivity," its "scientific" rationality, and its distancing from the moral stance of old-fashioned thinkers like John Ruskin. Green economics can and will promote the implementation of ecologically sound technologies, but it must also turn utopian economics on its head. Green economics must incorporate into its body of thought the dynamics of eutopian *culture*. A great deal of highly useful thinking has already been done, from Peter Kropotkin's *Fields, Factories, Workshops* to Harry Braverman's *Labor and Monopoly Capital.*

But there are dynamics of depth psychology (the psychoanalytical space that brushes up against the spiritual) operating in our present crisis. Culturally and spiritually, we are about to find ourselves in a vast historic swing back *toward* nature. We need to study how this ties in to the dynamics of gender, sexual identity, and sexual symbolism in economics. We need to look again at how sexuality is identified with nurture and with nature.

NOTES

1. Leopold, *Sand*, 261.
2. Braverman, *Labor*, 295.
3. Murphy, *Are*, 100.
4. Brutus, *Confessions*, 135.
5. *Columbia*, 1542.
6. Lefabvre, *Coming*, 131–32.
7. Phillips, *American*, 241.
8. Braverman, *Labor*, 193.

14

The Conscious Id

IN THE PREVIOUS ESSAYS, I attempted to deepen the analysis of, and find new meanings in, the forced contraction of rural culture. The shift from a predominantly agrarian society to an overwhelmingly urban society was seen to contain charged symbolic elements. These elements were, at the first level, nothing less than commonly accepted understandings: that horticulture and settled village life, for instance, could be traced to a feminine impulse, or (in the words of Thorstein Veblen in *The Theory of the Leisure Class*) that "Virtually the whole range of industrial employments is an outgrowth of what is classed as women's work in the primitive barbarian community."[1] But why insist that dry anthropological data are charged with "symbolic elements"? For the moment let's pursue an implicit answer to that question by attending to the "Male and Female Symbolism in the Thought of Ricardo, Malthus, Engels, and Marx"—a section title in Walter Weisskopf's book, *The Psychology of Economics*:

> In all manifestations of symbolic language [writes Weisskopf] we find a frequent identification of the earth, the woman, and the mother and a wealth of sexual symbolism in connection with agriculture and economics production . . . . In most cultures the earth has been identified with the woman . . . . The identification of heaven with the father and of earth with the mother is generally found in many mythologies . . . . In the primordial heritage of mankind, labour had a sexual and male connotation; we find the same connotations in the thought of political economists.[2]

As Weisskopf shows, the conception of whether land or labor was central to productivity shifted in economic thinking in a very short time. In the eighteenth century, the French physiocrats held that "all wealth originated with the land and that agriculture alone would increase and multiply wealth."[3] In *The Machine in the Garden*, a book on technology and the pastoral ideal, Leo Marx says that in the eighteenth century "Political economists and agricultural reformers now dwelled as never before upon the primacy of agriculture in creating the wealth of nations. While the physiocrats, extremists of the movement, insisted that husbandry was the *only* source of economic value, most of the experts, including the incomparable Adam Smith, agreed that agriculture was the primary and indispensable foundation of national prosperity." Marx is also aware that the "craze" (as he calls it) for the "charms of rural life" was largely provoked "by a recognition, however inchoate, of the impending threat to the supremacy of rural values"[4] embodied in the twin forces of industrialization and the Enclosure Acts. The machine was being wheeled into the garden.

By the time economic theory had passed through the mind and pen of Adam Smith and had gotten to Ricardo and Malthus, however, land was no longer held to be the primary source of wealth. Weisskopf says that for Malthus and Ricardo, land had assumed a "peculiar role"; it was "the source of all economic ills"; it had become an "evil principle" in their conception of economics.[5] Instead of pointing to land as the foundation of prosperity, they identified human labor—*male* labor in particular—as the real producer of value. This shift in economic value corresponds historically to the contraction of agrarian culture and the burgeoning of the new industrial system. Weisskopf also traces the development of Friedrich Engels' thought, based on the investigations of the American anthropologist Lewis Henry Morgan and the Swiss archeologist and myth interpreter Johann Bachofen, by showing that in Engels' conception of early human history the rise of private property (explicitly the ownership of land) emerged simultaneously with patriarchal monogamy. In Engels' analysis, according to Weisskopf, "The female is predominant in the family organization without and before the existence of private property; the male is the one who owns private property."[6] Once again, there are parallels with modern economic conditions: the eighteenth- and nineteenth-century enclosures in England forced an end to traditional patterns of communal agriculture by abolishing the commons and by evicting the small farm people and agricultural laborers, both women and men.

So in these revolutions in economic organization—the movement several millennia ago from Neolithic matrilineal society to patrilineal civilization, and the sudden shift in the eighteenth and nineteenth centuries from (female) land as the source of wealth to (male) labor as the source of value—the underlying dynamics are indeed charged with symbolic meanings that have concrete ramifications for the shape and tone of contemporary society as a whole. In *The Psychology of Economics*, Weisskopf says that Marx and Engels

> . . . identify the existing economic political and family system with the male father, against whom they revolt. To the oppressed they attribute a female character. We contend that a similar unconscious symbolism is at work in the thought of [Malthus and Ricardo]: they are against the female element, land and earth, because they side with the existing economic system, which has a male connotation.[7]

Yet, as Weisskopf goes on to point out, the "activistic ideal of Western economic civilization, which formed an essential part of the new economic value complex, acquired in the minds of contemporaries a masculine connotation; activity and masculinity became synonymous."[8] Perhaps we need only add that the "activistic ideal" of Western civilization always had a masculine connotation and that the industrial revolution merely permitted the systematic imposition of a masculine mechanical utopia.

Norman O. Brown, in his *Life Against Death: The Psychoanalytical Meaning of History*, attempts to tie this association between economic valuation and masculine activism into depth psychology:

> The psychology of giving is intimately feminine; the psychology of possession and taking is masculine . . . . Identification with the father is a way of denying dependence on the mother . . . . 'Taking' is a denial of dependence and thus transforms the guilt of indebtedness into aggression; and the masculine complex, the obsessive denial of femininity, is inherently aggressive.[9]

Well, let's take a breather from all these references, allusions, and quotations and see where we have come. The Earth has traditionally been identified with the female: Mother Nature, Mother Earth. In traditional preagricultural societies, food was simply gathered or hunted; it was not *produced* through human labor. Over time, women developed the foundations of agriculture through the discovery of horticulture via gathering. In the process, they provided the social impetus for the establishment of stable and settled villages, abundant food precluding the need for a nomadic life. These villages grew in size and became cities. But somewhere in the early period of the expanding and specializing city, women's sphere was forced to contract. Especially with the development of the plow, men took to (and over) the fields. Matrilineal values were shrinking; patrilineal values were expanding. Yet the city is something more than an overgrown village. What it reflects, says Norman O. Brown, is

> . . . the new masculine aggressive psychology of revolt against the female principles of dependence and nature . . . . In the new space of the city, which is always a sacred space, man succeeds for the first time in constructing a new life which is wholly superfluous, and wholly sacred. A city is by definition divorced from primary food production, and therefore by definition superfluous; its whole economy is based on the economic surplus.[10]

Brown also says that the "alienated consciousness [of the civilized mind] is correlative with a money economy. Its root is the compulsion to work."[11] He then goes on to say that "Money is the heart of the new accumulation complex; the capacity of money to bear interest is its energy; its body is that fundamental institution of civilized man, the city."[12] So if land, agriculture, and nature generally are all identified with the female in traditional symbolism, then the new sacred city, with its aggressive economic dynamism and interest-bearing money, is associated with the male. On this point—the nature of money—there is a good deal of confusion in psychoanalytic theory. The first level of observation, of course, is that Freud and his closest associates tended to identify money with feces. So Brown can say that "Money inherits the infantile magic of excrement and then is able to breed and have children: interest is an increment."[13] But feces do not breed; *semen* does. Money is the sperm of utopia. Referring to the studies of the economic anthropologist M. J. Herskovits, Brown says that

> [Herskovits] connects the division of labor with economic surplus, and connects economic surplus with prestige and privilege, and, as we have seen, connects prestige and privilege with the domain of the sacred. And it is easy to show a continuous development of the division of labor as an integral part of the expanding sector of the sacred-superfluous, climaxing in the elaborate organization of the divine household by the priest-kings of the earliest civilizations.[14]

This "elaborate organization" of the "expanding sector of the sacred-superfluous" has now been "democratized" in the technocratic affluence of the utopian suburb. In line with his remark that the root of alienated consciousness lies in the compulsion to work, Brown also says that "This compulsion

... subordinates man to things, producing at the same time confusion in the valuation of things and devaluation of the human body."[15]

Weisskopf describes in greater detail the social context in which this confusion in the valuation of things emerged:

> The emergence of the market economy was accompanied by the destruction of social ties and bonds. Large numbers of small peasants, yeomen, cottagers, etc., were uprooted from the soil and driven into the factory towns, where they had to earn their living as labourers. The family structure and the character of family relations were changed. Mobility made permanent, lasting ties more and more difficult. The growing division of labour severed the relationship between the worker and his product and prevented him from finding satisfaction in creation. The market economy tended to disintegrate social relations and to destroy the traditional habitat of people. The industrial system functionalized man, alienated him from himself, and depersonalized his relationships with others. The attitude of competitive acquisitiveness, required and enforced by the market system, is in conflict with the need for primary, warm, close, affectionate personal relations. The arguments of economic theory that competition is economically beneficial will not eliminate the fact that the individualistic, competitive value-attitude system led to aggressive interpersonal relations. This hostile attitude toward others, originating in the market place, permeated all spheres of social and human relations. It exercised a disintegrating influence on all primary social and human ties of solidarity and community.[16]

Weisskopf then goes on to make the critically important point that these "social and human ties" are connected deeply and strongly with a symbolism that is feminine in its essence:

> Industrial man was deprived of many traditional social ties and human bonds which were in conflict with the newly emerging socio-economic system. He became isolated and separated from the protection which his group-membership had given him before. He reacted to this experience by identifying the previously existing community and solidarity with his own ties to his mother and to the female in general. In his unconscious and in his life-history, the archetype of separation from community was the separation from the mother.[17]

II

Our polarity here has been the feminine Earth, on the one hand, and the masculine heavens, on the other. Land, nature, natural productivity, agriculture, the early agrarian village, communal solidarity, and women are all to be seen on the female end of the spectrum. Labor, industrial techniques, money, the "sacred" sphere of the city, individualism, competitive values, and men are all placed on the male side. But let's shift our emphasis slightly and proceed with a tentative "application" of three Freudian concepts: the id, the ego, and the superego. If we are to attempt to correlate the psychoanalytical components of civilized "man" to the outline of utopian civilization—the id as the fundamental mass of life tendencies, the ego as the conscious and self-preserving self, and the superego as the concentrated imperatives of religious conviction—we would, for the sake of intellectual consistency, put the id on the feminine end of the spectrum and the superego on the masculine. In its most elemental form, the id is living nature, guided by the "pleasure principle" of feeling, shaping itself organically into a coherent whole (e.g., the vegetative community) without recourse to conscious morality, as the term is ordinarily understood. The superego, on the other hand, is the crystal out of which the "reality principle" of judgment and self-denial emerges. The id does not operate on the basis of self-denial; rather, though capable of enduring extended delay or sleep-like dormancy, it seeks organic fulfillment. The superego strives to conform to the dictates of supernatural law; it is inherently "sacred," in the transcendent meaning of the word.

In other words, the id is symbolically represented in reality by nature, by natural productivity, by land, by women, by organic agriculture, by the early agrarian village, and by the peasantry. ("Paganism" has typically been associated with conceptions of fertility antithetical to the ascetic monotheism of urban-based Father religions. The words "pagan" and "peasant" derive from the same Latin root, *pagus,* meaning "country district." The id, therefore, represents "pagan" energies of cultural "backwardness," energies that utopian civilization harnesses or extirpates, depending on historical circumstances.) By contrast, the stern dictates of the superego are embedded in authoritarian male power, in the laws and commandments of the "fathers." (Consider: the Fatherhood of God, the religious fathers, the founding fathers, the city fathers.) The city, then, as Norman O. Brown has said, "reflects the new masculine aggressive psychology of revolt against the female principles of dependence and nature." Marcus Borg, in *The God We Never Knew*, says the monarchical model of God "reinforces the superego" and "easily confuses God with the superego," and that this is a "central dimension of the human predicament from which we need deliverance."[18] The superego strives to control the id in a manner that parallels the ways in which the city exerts control over the countryside, how civilization controls agriculture. But where is the ego in this analysis? If we are to identify the superego with a group or class, it would be the ruling elite: the king, the priest, the banker and financier, the military leader. The id, in turn, would be women generally; but as a social class, the id would correspond to the peasantry. That leaves the middle class as the ego, urban (or suburban) in lifestyle and living standard if not also by residence.

Finally in psychoanalytical theory, it is libido that energizes the organism; and libido is defined by *Webster's* as the energy, motive force, desire or striving, either as derived from sexual drives or from the primal urge to live. In Lewis Mumford's *The Transformations of Man*, we read that

> . . . civilization, at least up to our own time, has always been sustained by pagan vitalities, undivided and undepressed; and when these ceased to be sufficient, it was redeemed again, as we shall see, by the opening up of a post-civilized prospect . . . .[19]

We can once again make a suggestive correlation: the "personal" libido of the individual organism corresponds to the collective energy of "pagan vitalities" within the historical context of the utopian city. The "libido" of civilization is located in the unsophisticated energy and undepressed exuberance of the peasantry and indigenous peoples. This is the human force that throughout the history of civilization has renewed the life of the city. Just as the city demands a constant influx of natural materials (the most basic being food), so, too, does it require its human energy to be replenished by the uncivilized vitalities of the *pagus*.

III

With these psychoanalytical correlations in mind, let us proceed into a speculative analysis of what Mumford calls a "post-civilized prospect." We are back again in political theory and its underlying symbolism. The ultimate vision of both Marxian and eutopian socialism is embodied in the concept of the withered state. Because the state is the center of political control, and because its center is the core of the (masculine) city, and because we have already identified this urban governing core with the superego, we can rather easily associate the withering of the state with the withering of the superego. At least we can say that the former clearly implies the latter. (Here we have a fuller insight into why Isaac Asimov's "ready-made ruling class" language is such a disaster. A ruling class is in control of the spiritual and cultural superego, and, as Marcus Borg says, this is a "central dimension" from which we "need deliverance.")

In his *Transformations of Man*, Lewis Mumford raises an important and (from a revolutionary perspective) disturbing question:

> For those who mastered the arts of civilization, it brought, no doubt, an intensification of life, a heightening of consciousness, a sense of the individual ego combined with a pride in collective achievement that justified the sacrifice it demanded. But those upon whom the sacrifices were imposed formed a majority of the population, and their willingness to submit to this compulsive routine remains, even now, a little mysterious.[20]

Perhaps the "willingness to submit" of the exploited class is no more mysterious than the willingness of nature to "submit" to the rapacious exploitation of *homo economicus*. Nature has always suffered her losses, apparently, without a sense of vengeance or retribution: grass grows to the stone wall and stops; it does not develop conscious strategies to overpower the rock and mortar. Unless we interpret massive natural phenomena such as hurricanes, volcanic eruptions, earthquakes, tidal waves, droughts, and the like, as expressions of nature's blind and impersonal rage (as a provoked animal will not necessarily single out the provoker for attack), we are left with the impression of the sustained passivity of nature. That is, the id is forever seeking to fulfill itself through natural processes of growth and consummation, but it does so blindly, intuitively, without recourse to rational strategies. Nature is "passive" in the sense that it endures without active resistance. In theological terms, the force of nature's id is immanent, indwelling; the force of the historical, civilized, utopian superego is transcendent, above and outside of. The contemporary Christian Right in this country, with its explicit and aggressive conception of God as male, judgmental, fierce and transcendent, now constitutes the "domestic" base for the increasingly raw and brutal utopian superego of civilized governance. And this is, as Marcus Borg has said, the "central dimension of the human [and global ecological] predicament from which we need deliverance."

If, therefore, the post-industrial age or (in Mumford's terminology) the "post-civilized prospect" is to emerge from the contraction of male-controlled and male-directed civilization, there would seem to be an implied shift from the transcendent "theology" of the superego to the immanent "theology" of the id. By "theology" I mean only incidentally the productive contributions of theologians; what is meant by the word in this context is a *Weltanschauung*, a world-view, an ethos, an understanding of the divine as compassionately *internal* to life's unfolding rather than judgmentally *external*. Clearly, the emergence of the women's movement in advanced industrial society, as well as growing ecological awareness and even the symbolic meaning of the recent death-of-God theology, implies an impending radical shift in our understanding of ourselves as cultural beings, and a re-evaluation of our place and function within nature. "The future distribution of energy and resources among communities should be based on the integration of human and natural ecosystems," writes Carolyn Merchant in *The Death of Nature*: "Such a restructuring of priorities may be crucial if people and nature are to survive."[21]

But how is this "restructuring of priorities" to be effected? The answer at the first level is increasingly clear: the urban-industrial system has surpassed the "limits to growth." Ecological viability at this point in the history of civilization demands the contraction of the expansive superego and a reorientation toward the nurturing id. The deeper question is whether the necessary transformation can and will be achieved with a minimum of violence and systemic breakdown, or whether resistance to change will block transition and thereby make violence and systemic breakdown inevitable. Where is the modern ego in all this? We have already identified the ego with the city's ordinary inhabitants. In the Marxian scenario, the industrial working class was to bring about the liberating revolution by overthrowing the superego of the urban elite and then instituting, before the withering of the state, a dictatorship of the proletariat. In other words, it is the alienated and exploited men who *make* the revolution just as their productive energies are reflected in the labor theory of value.

In referring to Marx's development of the labor theory of value, Walter Weisskopf says that

> . . . Marx's idea that value is determined by labour bestowed in production and not by the exchange process means, unconsciously, that only the producing active male procreates and that

> the woman has nothing to do with it. The role of the female is unconsciously belittled.[22]

Carolyn Merchant traces these sex-based ideas back as far as Aristotle where "Power and motion were contributed solely by the semen."[23] The revolutionary sons turn against the reactionary father and overthrow him; but they do so from within the intellectual exigencies (e.g., the labor theory of value) of the father's world-view. Male workers, like soldiers in the "father's" army, have been systematically "disciplined" in the rigorous brutalization of esthetic and ethical sensibilities. That this is true for soldiers requires no explanation, for soldiers *embody* the punitive and controlling dictates of the urban superego; they are the regimented fingers and toes of the utopian phalanx. The destruction of esthetic and ethical coherence for the mass of industrial workers, however, is inherent in the very processes of mass production. Only in the remotest sense can a worker feel that the resulting product expresses either his or her creative impulse. In addition, the workplace environment is designed, built, owned, and managed by economic forces—forces that can be called human only in the remotest sense—quite beyond the influence of the worker. The worker is a mere cog, an integer in Personnel. The esthetic and ethical conditions necessary for organic self-regulation have no place in such a schema geared only toward maximizing productivity, efficiency, and profits. (For an exceptionally lucid analysis of the plight of the working class in technological society, see Harry Braverman's *Labor and Monopoly Capital: The Degradation of Work in the Twentieth Century*.)

It is, therefore, not terribly surprising that, once in command, the alienated male proletariat, lacking eutopian cultural memory and trained in utopian efficiency, should indulge in the punitive tendencies of the superego. Industrial ideology is saturated with utopian expectations as projected through political organization and technological infrastructure. Herbert Marcuse, in his *Eros and Civilization*, shows us a crucial tipping point where utopian consciousness takes command:

> Whereas the ego was formerly guided and driven by the *whole* of its mental energy, it is now to be guided only by that part of it which conforms to the reality principle. This part and this part alone is to set the objectives, norms, and values of the ego; as *reason* it becomes the sole repository of judgement, truth, rationality; it decides what is useful and useless, good and evil. *Phantasy* as a separate mental process is born and at the same time left behind

> by the organization of the pleasure ego into the reality ego. Reason prevails: it becomes unpleasant but useful and correct; phantasy remains pleasant but becomes useless, untrue—a mere play, day-dreaming. As such, it continues to speak the language of the pleasure principle, of freedom from repression, of uninhibited desire and gratification—but reality proceeds according to the laws of reason, no longer committed to the dream language.[24]

Utopian organization, through such agencies as economic structure, religious ideology and educational institutions, has gridworked society with its "laws of reason," and the social ego (i.e., the working class) has internalized the punitive and control-oriented predilections of the utopian superego—hardhats with flag decals and fundamentalist churches with lurid apocalyptic fantasies, all of it goaded into continual seething resentment and misdirected rage by manipulative AM radio propagandists. A distant "withering" of these predilections is envisioned in the Marxian schema, but the intervening steps that lead toward and bring about such "withering" are pretty poorly delineated. (They are poorly delineated because a withered state presupposes a transformation of consciousness from utopian to eutopian, from transcendent to immanent, and utopian consciousness declines to let go of itself.)

In essence, the economic and military soldiers (from the Latin *solidus,* meaning a piece of money; hence soldier's pay; hence, given the suggestive thoughts on the origin and nature of money, seminal dregs) are not at all eager to relinquish their control and (male) prerogatives. The dictatorship of the utopian proletariat is therefore a logical development of resentful alienation in which the power inclinations of the repressive superego are introjected and maintained by revolutionary sons. Now the eutopian, Green, and anarchical socialist perspectives place a different emphasis on this analysis. Most importantly, there is to be no dictatorship of the proletariat. The male industrial workers and their military counterparts are not, in the eutopian schema, blessed with the status of a new elite. Rather, there is a basic trust in the cultural intelligence of the id, an affirmation of nature and the Tao. Eutopian-anarchist literature (e.g., the works of Murray Bookchin, E. F. Schumacher, Paul Goodman, Martin Buber, Peter Kropotkin) points toward an *immediate* withering of the state or superego and a reconstruction of natural values.

In the Marxian scenario, the superego is overthrown by the alienated (male) ego but the underlying utopian (anti-female) values of the superego are ingested and maintained. In the eutopian, Green, and anarchist scenario, the superego is *dissolved* by the refusal of the ego (both male and female) to be continually polarized by the utopian agenda. But the contraction of the superego (withering of the state) clearly implies the curtailment of urban domination and the revitalization of rural culture, and it is precisely this fear of "backwardness," this clinging to utopian civilization, that is the stumbling block for both mainstream socialism and mainstream feminism. The classical superego has so inculcated the ego with fear of the id that the response of the ego to the destruction of rural culture was one of numb indifference. By long utopian training, the ego has come to perceive the id as insatiable and chaotic, destructive, violent, lustful without limit, and innately stupid. In reality, the id operates on a pendulum of need and fulfillment, is only blindly violent, and is forever creating eolithic esthetic harmonies. The id is also almost entirely helpless in defending itself against an entrenched ego empowered by the stern utopian dictates of the superego. The ally of the peasant id is nature; the ally of the urban ego is civilization, *civis,* the city.

The contemporary women's movement is certainly one of the most important measures of the transformation from the transcendent superego to the immanent id. The women's movement in many ways strikes at the symbolic as well as the functional root of the utopian agenda. The ecology and back-to-the-land movements are two other expressions of this transformation. These movements are all charged with symbolic content; and if we can become conscious of these elements and permit this new consciousness to function more deeply in our lives, our actions in behalf of the Earth, Green culture, gender reconciliation, and Green politics will gain a much keener clarity and effectiveness. As Carolyn Merchant has pointed out:

> The ancient identity of nature as a nurturing mother links women's history with the history of the environment and ecological change. The female earth was central to the organic cosmology that was undermined by the Scientific Revolution and the rise of a market-oriented culture in early modern Europe. The ecological movement has reawakened interest in the values and concepts associated historically with the premodern organic world. The ecological model and its associated ethics make possible a fresh and critical interpretation of the rise of modern science in the crucial period when our cosmos ceased to be viewed as an organism and became instead a machine.[25]

The changing social ego we are all part of in our historical time must be drawn into the service of the natural id. The effort, as Norman O. Brown has repeatedly insisted, is to make the unconscious conscious, to bring the life-affirming impulses of the id into the ego, to eroticize in the broadest sense of the word our interactions with "objective" reality, and to transform agribusiness into organic gardening and industry into craft. We must become, in many important respects, the conscious id. Yet it may not be enough to postulate, merely, a restoration of the life-affirming id. Nor is it finally appropriate to suggest that the superego is likely either to completely wither or dissolve. The issue is not to "dissolve" civilization but to reduce its scale, accent its culturally useful features, and bring it into balance with folk culture. Herbert Marcuse, in *An Essay on Liberation*, calls for the "transfer of moral standards . . . from the Establishment to the revolt against it." Marcuse sees that to turn from the "institutionalized fathers" toward a free and ecological society requires that the "vested interests in the existing system . . . fostered in the instinctual structure of the exploited" must be overcome. He wishes to activate

> . . . the elementary, organic foundation of morality in the human being. Prior to all ethical behavior in accordance with specific social standards, prior to all ideological expression, morality is a 'disposition' of the organism, perhaps rooted in the erotic drive to counter aggressiveness, to create and preserve 'ever greater unities' of life. We would then have, this side of all 'values,' an instinctual foundation for solidarity among human beings . . . .[26]

Neither the continued repression of the peasantry/id nor the catastrophic collapse of civilization/superego is a viable solution to the human dilemma. What's needed is an integration of countryside and city, an integration of human personality: a balance of elements rather than a mechanical formula or ideological tinkering.

Yet the *hegemony* of father images must wither or dissolve. This is a process already going on, working its way through cultural self-awareness. The enlargement of the public sphere of women—social, political, economic, religious—reveals this to be true. It is ironic that Western civilization, with its dynamic masculine roots in both the Judeo-Christian ethos and Greco-Roman rationality, should be the very historical medium through which father images have reached their obsolescence. The exhaustion of global frontier, the development of instantaneous intercultural communication and rapid travel, the rise of standardized multinational commodities, the breakdown of regional, communal and folk identities, the standardization of prepackaged educational curricula, and the ultimate threat of extinction by atomic weapons, have brought the entire world to the brink of holocaust or the cusp of transformation. The control-oriented superego, shaped and concentrated by transcendent images of stern patriarchal divinity, has created the objective global conditions that demand the contraction of utopia. Either contraction or catastrophe: the choice is now a real one: integration or disintegration, a new stability or an unimaginable chaos.

The revival of feminine values suggests not only the feminization of (a more modest) civilization but, through symbolic association, a reassertion of Land (not Labor, not Capital) as primary in economic theory. Green politics, even to the symbolism of its chosen color, is the politics of Land, of gardening, of Earth, of nature, and of a dynamically creative ecological culture. Eutopia implies the feminization of the superego.

NOTES

1. Veblen, *Theory*, 23.
2. Weisskopf, *Psychology*, 120, 121, 123–4, 145.
3. *Columbia*, 1542.
4. Marx, *Machine*, 98–99.
5. Weisskopf, *Psychology*, 125.
6. Weisskopf, *Psychology*, 130.
7. Weisskopf, *Psychology*, 131–32.
8. Weisskopf, *Psychology*, 136.
9. Brown, *Life*, 280.
10. Brown, *Life*, 281–82.
11. Brown, *Life*, 237.
12. Brown, *Life*, 281.
13. Brown, *Life*, 279.
14. Brown, *Life*, 261.
15. Brown, *Life*, 237.
16. Weisskopf, *Psychology*, 140–41.
17. Weisskopf, *Psychology*, 141.
18. Borg, *God*, 66–67.
19. Mumford, *Transformations*, 55.
20. Mumford, *Transformations*, 45.
21. Merchant, *Death*, 295.
22. Weisskopf, *Psychology*, 147.
23. Merchant, *Death*, 13.
24. Marcuse, *Eros*, 141–42.
25. Merchant, *Death*, xx.
26. Marcuse, *Essay*, 8–16.

15

Toppling the Sexual Avengers

I SUBSCRIBE TO *THE Nation*, a more-or-less weekly left-of-center news magazine out of New York City. (For ethical consistency, I like *The Progressive* more, but *The Nation* covers a larger range of topics.) Once in a while an issue arrives that just brims with energized content. *The Nation* for March 24, 2003, is such a one. It has a very focused snapshot of Kurdish troubles by Charles Glass, a similar picture of the Korean crisis by Bruce Cumings, a critique of the *Washington Post*'s rightward drift by William Greider, an article on Laura Bush and Iraq (disguised as a scene from a play) by Tony Kushner, a women's sports analysis by Ruth Conniff, outstanding book reviews by Ian Lustick and Dusko Doder, and—I won't cite everything—an article clumsily entitled but cleanly written by Richard Goldstein, "Neo-Macho Man: Pop Culture and Post-9/11 Politics."

It's the Goldstein piece that prompts me to put pen to paper, though virtually all the other articles in *The Nation* could be read as supporting evidence for Goldstein's thesis. His effort is to explain the

> . . . reaction to feminism that began in the 1980s and advanced in the '90s, even as the empowerment of women became a tenet of Democratic politics. As women rose, so did male anxiety, and in this edgy climate a new archetype appeared in pop culture: the sexual avenger. His rage often focused on personal betrayal, but implicit in his tirades was a sense of the world turned upside down.

"Sexual avenger" is a key term in this "backlash culture," this "flight from feminism," this "thug package" in the "broader culture of misogyny." Hip-hop celebrities figure in this analysis, including white rapper Eminem. But, according to Goldstein, what really enabled this culture of misogyny to wildly inflate were the terrorist attacks on September 11: "In its wake, the once-mocked figure of the dominant male has become a real-life hero," with a "new spirit of patriarchal vitality," including Donald Rumsfeld with his "feckless swagger" and "timeless union of militarism and macho." Goldstein says, by way of contrast, that the Depression-era policies of Franklin Roosevelt weren't macho, because FDR "arose from a culture that regarded protecting the weak as an important manly virtue." But the "dark side of macho" has now turned against "strong women and weak men":

> Male grievance has found a geopolitical target in Saddam. Sexual revenge has been sublimated into military payback. Underlying this process is a sense of the world as jungle where friendship is transient, danger is everywhere and one can never have enough power. This is the classic rationale for macho. Feminism teaches us that it's a pretext for preserving the order. Liberalism tells us it's paranoid. But what once seemed like paranoia is regarded as reason, and what was piggy now feels natural.

Immediately Goldstein thrusts Eminem forward ("this monster from the id") as a model of piggy macho—lyrics about raping his own mother, about "slaughtering every bitch in sight." (Elvis Presley, says Goldstein, "embodied a different morality than Eminem does. His appeal was Dionysian rather than sadistic; his lewdness didn't preclude the possibility of love.") And, says Goldstein, referring to Eminem's "fictionalized biopic [*8 Mile*] set in streets so mean that even the sun stays out of sight,"—

> It's no coincidence that *8 Mile* ruled the box office right after Bush's GOP romped at the polls. These two young patriarchs seem utterly opposite, but they have fundamental things in common. Both are social conservatives who stand for a male-dominated order. Both owe their appeal to anxiety over sexual and social change. Both offer the spectacle of an aggrieved man reacting with righteous rage. These qualities, which once seemed dangerous, now read as reassuring. The macho stance that once looked stylized is now a mark of authenticity.[1]

There's more here, which we'll get to in a moment, but first it's necessary to attend to Goldstein's phrase "this monster from the id." While I am no authority on the plumbing of psychoanalytical drainage systems, it seems to me that "id" in Goldstein's usage corresponds pretty tightly to "sexual avenger," and both terms fit snugly into the political psychodynamics of "social conservatives who stand for a male-dominated order," for "aggrieved [men] reacting with righteous rage." Excuse me if I'm wrong, but I'm under the impression that such an order is closely allied not with the id but with the superego. This is Fatherland. This is Male Prerogative. This is the Omnipotent Fatherhood of God.

This is also, on Goldstein's part (I presume unwittingly), a term usage—"this monster from the id"—which is a familiar, totally conventional way to scapegoat the unconscious. But, as we have seen in the previous essay, to scapegoat the id is, by association, to scapegoat the feminine, the primitive, the "backward," and the rural. It is to heap all these in a negative semantic pit. And it is precisely this tendency, operating largely beneath consciousness, that disenables the Left from achieving a deeper and stronger critique of civilization. Afraid of the id, the Left reflexively seeks the protection of the superego even as it struggles against policies and practices (e.g., the Vietnam War, the ongoing debacle in Iraq) that are manifestations of superego brutality—which very brutality is then misidentified as monstrous id eruptions. Thus "civility" walks unscathed from every disaster and is accorded the veneration of a hero or a saint. Threaded through all this, even in the mental structure of self-described atheists, is an underlying fabric woven by the monarchical model of God, a reflexive clinging to transcendence and a corresponding fear of the immanent.

It's not just the Right that is enmeshed in utopian mentality. We all live within the utopian *system*. Utopia has shaped every single one of us to some degree. Getting a grip on the magnitude of this shaping requires moving outside of, or beneath, utopian mentality. Learning how to do this, particularly as authentic alternative cultural formations disappear, becomes increasingly a matter of eutopian spiritual practice, in which sustained, disciplined meditation is a key element, and in which Spirit (with or without a capital S) is recognized as internal to life's unfolding rather than judgmentally external. This will enable us, strengthen us, to develop clear and loving eutopian alternatives to the globally ruinous utopian hegemony. But without such internal clarity we will be unable to adequately disengage from the utopian mental constructs that hold us in their grip, thereby deflecting critical analysis from victimizer to victim, from superego to id.

Two paragraphs later, without saying so, Goldstein partly plugs this psychopolitical leak when he says that

> It's easy for Republicans to seem manly, for the same reason pundits call the GOP the Daddy Party. Their tough-love style represents

> patriarchal values of strength and order. If the Democrats are (often disparagingly) called the Mommy Party, it's because their attitude expresses feminist values of empathy and equity. Democratic men are not less masculine than Republicans, but they tend to be less macho in their manner, reflecting an etiquette that allows both sexes to project power. This is also why Democratic women tend to be less courtly and decorated than the daughters of the GOP. When voters see these qualities in a candidate, they are reminded of the underlying sexual politics. If Democratic men seem weak and Democratic women all too strong, it has much less to do with character than with the angst that the party of feminism generates.

Gender, says Goldstein, "is the great unmentionable in public life, and women are especially invisible as citizens in a time of crisis":

> It's even harder to address the culture that animates these policies. No progressive wants to be a censor, a puritan or, worse still, a fogy. But attention must be paid, because cultural values are central to social reality. A norm can only be undone if people understand the damage it does, and macho is a stunting force even when it looks fresh and young. Under its thumb, a generation is growing up with attitudes that will warp their lives, not to mention the course of American politics.

This "primal issue" of the macho code is rarely broached, says Goldstein. Which means that deference to male authority, fear of male attack, remains largely operative as the underlying form of social, economic, and political control. What will it take, Goldstein asks, to address the "relationship between male dominance and the current crisis?" Since the real issue of sexual equity will not go away and will, in fact, intensify, the "only option for the Mommy Party is to embrace its identity. That means stripping Republican macho of its mystique."[2]

Whatever the complex relationship between id and superego (I use the terms because they're part of a Freudian vocabulary that still contains dynamic insightful content), it should be clear—even as we identify women, historically, with id and men, historically, with superego—that what is loosely called women's liberation has resulted in women entering, in ever larger numbers, the professions, fields of authority and expertise formerly held solely by men, especially medicine, law, politics, and religion. Insofar as it is credible to identify these governing professions with civilized superego—and I see no reason not to do so—it's obvious that women are, as a gender group, accelerating their collective entrance into superego governance. Women are rising toward a feminist reformulation of the cultural superego. The traditional male superego, clinging to its prerogatives, enormously buttressed not only by the history of conventional male behavior but also by religious mythology that threatens eternal hellfire to gender apostates, both creates and utilizes crises to intensify and reconfigure its grip on power, its magnitude of control. Thus the George W. Bush administration throve on crises (preferably inflated ones), invariably demanding additional authority to combat yet another layer of perceived threat. (One is struck by the political parallels to apartheid South Africa, as unfolded in *Nelson Mandela* by Martin Meredith.) The Bush doctrine was one of perpetual crisis met with suppression and force.

Insofar as id, in formal psychoanalytical theory, represents the fundamental mass of life tendencies and *not* some dreadful monster energy—the id is fundamentally a positive, not a negative, construct—one can only hope that as women invade the "sacred" sphere of the traditional male superego, they will bring with them into a newly reconfiguring feminine superego a huge load of "life tendencies." While it's true that human life cannot proceed without judgment—yes to this, no to that—the quality of human judgment is heavily determined by the magnitude of mercy and compassion contained within the judging temperament. Conventional male authority has always been short on compassion and mercy, its fuse stubby, its demeanor harsh, and its capacity for violence imminent. Cumulative male authority, intensified beyond ecological sustainability by science, technology and industry in these last centuries of utopian mushrooming, has reached—has *produced*—a global crisis that demands a fundamental transformation of human attitude and conduct. The "sexual avengers" who have proliferated since the Reagan years represent an explicit male fascism of utopian reaction. September 11 served to tighten the noose. Global forms of contending reaction now accelerate disaster. Violence is the common currency of both change and order. This process will continue until male exhaustion is reached or Mommy Power puts a stop to it—or some combination of exhaustion and outrage topples the sexual avengers.

At many levels, the forces of reaction have already lost. The party of feminism (whether or not it's the *Democratic* Mommy Party) has already risen, like the African National Congress of the 1950s and '60s, to an inevitable victory. It's now a question of how much destruction the avengers will inflict on the world before even they wake up from their nightmare of "avenging"—to discover that eutopia is infinitely more livable than utopia in its inevitable state of collapse.

NOTES

1. Goldstein, "Neo-Macho," 16–18.
2. Goldstein, "Neo-Macho," 18–19.

16

The Dangerous Female

In the January 26, 2004, issue of *The Nation*, Richard Goldstein was back ("Butching up for Victory") with "attitude," "sexual fantasies," "erotic aura," "gender presentation," and "the gender card." The subject was Howard Dean as "a manly alternative to Bush's ripe macho." Or, perhaps, the subject was whether a Democrat can "be an alpha male." The writing is all a bit self-consciously hip and precious, but there is real psychoanalytical probing going on beneath the verbal flash. There are several key insights in Goldstein's article: that Bush "owes his mandate, such as it is, to his projection of macho"; that, currently, "status conflicts are driven less by economic fears than by threats to the masculine mystique"; and that the answer to why we require "macho magic from our leaders" is "primal, and therefore complex."[1] Goldstein invokes George Lakoff, a Berkeley linguist, whose critique of how Republicans have successfully framed issues is a bit of a rage these days. Lakoff's key concepts seem to be that "the strict father model" constitutes the "conservative" world-view, and "a nurturant parent family" represents the "progressive" world-view.[2]

Professor Carol Burke, in the March 22, 2004, issue of *The Nation* ("Why They Love to Hate Her"), uses Jane Fonda as the men's urinal poster woman ("one vet recalled his pleasure at urinating on her face") to explore how "military culture" utilizes such images and notions to "stabilize and punish the dangerous female."[3] How do we deal with these flashy ideas and explicitly brutal images in a way that integrates them meaningfully into a deeper historical understanding? In my estimation, civilization has been energized by how the strict father has been able, so far, to successfully stabilize and punish the "dangerous female." But now we have unconditionally entered a world in which such predication is totally counterproductive and, to say the least, obsolescent. Having pursued "wilderness" with both space craft and computer-aided microscopes, having coerced the Earth into an astonishing new phase of species extinctions, our aggressive interrogation of nature is already facing the inevitable consequences of ecological blowback—which, in turn, are "managed" with much of the same haughty arrogance that has brought us to this predicament.

The World's Only Remaining SuperPower has never been truly defeated, does not know what it means to be catastrophically helpless and at the mercy of stronger forces. If one can close one's eyes and imagine, in a sort of speeded-up movie kind of way, the amazing expansion of Euro-American power from its tiny toeholds on the eastern seaboard of North America in the seventeenth century to its current global supremacy, one can begin to appreciate the depth and power of the "strict father" arrogance that has been the guiding rudder of this penetration. It was and is a "Manifest Destiny."

Such a world-view, such a theology, such a character structure and mass psychology, has enormous temporal momentum. That we conventionally call it "conservative" correctly implies that it has ancient roots and lineage even as it *conserves* nothing but its own material advantages and fanatical self-obsession. This is a beast of enormous wallowing magnitude. It contains the core thrust of ancient and archaic civilization with all the self-evident righteousness and oblivious superiority that has blinkered every known civilization through the ages. Our civilization—this one—may be the most unself-reflective of them all: The Greatest Country That Has Ever Existed In The History Of The World. Self-reflectiveness is, in fact, taboo. Anyone, for instance, who attempts to seriously examine the underlying causes and motivations within contemporary "terrorism" will be summarily treated as a heretic, a subversive, an enemy of civilization. We are prevented by social pressure, mass consensus, conventional politics, compliant religiosity, and the corporate media from exploring (much less deconstructing) the impending deadly consequences of our structural obsession.

Republicans seem truly locked into this obsession. They are, in the main, True Believers. But Democrats represent a more muddled picture. Just to fabricate numbers out of my imagination, I would say Democrats are sixty percent True Believers, ten percent Alarmed Radicals, and thirty percent Twilight Zone. Afraid to *lose* elections, Democrats constantly adjust sails to the prevailing wind and thus perpetuate the dominant mythology. And, since the dominant mythology is speeding us obsessively toward a cornucopia of disasters, this juggernaut is not being either identified or challenged, except by a loosely organized secular Left (let's call it the Green Chomsky/Zinn Left) and the religious Left (the Catholic Worker movement, above all, but also the Quakers).

E. Badian, in *Roman Imperialism in the Late Republic*, said that "The obedience of the weak to the strong was, to the Roman aristocrat, nothing less than an eternal moral law."[4] Obedience to this "eternal moral law" in this century will bring a horrific magnitude of ecological and cultural disaster. The measure of our aptitude for either preventing, evading, or recovering from these catastrophes is the degree to which The Dangerous Female is capable of shaping a new cultural sensibility.

NOTES

1. Goldstein, "Butching," 11–13.
2. Lakoff, *Don't*, 6.
3. Burke, "Why," 16.
4. Badian, *Roman*, 15.

17

Masculine Politics

To affirm the need for a feminine superego we are obliged to examine the masculine superego in its most visible form—politics. And as the utopian political machine finds itself increasingly bogged down in accrued contradiction—most fundamentally that the utopian standard of living is ecologically unsustainable—the denial of these contradictions by conventional politicians shows the negative side of the masculine superego in a stark and garish light.

As I worked on a draft of this essay in the late summer of 1988, vice-presidential candidate Dan Quayle was under attack because of allegations that his family's wealth and political influence enabled him to join the Indiana National Guard (ahead of others on a waiting list) in order to avoid regular military service during the Vietnam War; presidential candidate Bush was defending Quayle for not dodging the draft by going to Canada, destroying his draft card, or (images of hell!) burning the flag; presidential candidate Dukakis was under attack for vetoing a bill that would have made the Pledge of Allegiance mandatory in the Massachusetts public schools; and vice-presidential candidate Bentsen acted as Dukakis's shield against the Right because Bentsen was as far to the Right as Bush and Quayle. All this deference to the Right means that the utopian superego is still the force to be reckoned with at the level of public policy and dominant male imagery: heroic male competition in an effort to show who is most patriotic and who has "balls" enough to run the utopian system. (George Bush's activities in the Central Intelligence Agency were so sacrosanct they couldn't even be alluded to in what passed for political "debate.") And, since conventional politics is utopian theater, it's appropriate that we turn our attention to a movie.

In St. Louis, in the early winter of 1974, a friend and I went to see *The Godfather*. Near the end of the film there was a scene that, judging by the reaction of some men in the audience, proved to be the most dramatic. It was not a murder, nor was it the enactment of any sensational crime. It was a simple pause followed by the word "no." The situation developed as follows. The "godfather" dies and the son who earlier in the film had been portrayed as a sensitive, nonviolent sort of fellow assumes control of the Mafia "family." This son has married an intelligent woman from outside the Mafia families who knows nothing about her husband's activities as a mobster. When the evidence begins to build and her suspicions grow, she finally confronts her husband with questions about his "business." At last he appears to relent, allowing her "this one time only" to ask questions about his "business." Her questions have to do with his mob activities, especially the disappearance of a young man from within their own family. Her question in essence is: "Are you a part of this terrible and violent thing, the Mafia?" Every viewer already knows the answer to that question is a capital-Y Yes; but the young "godfather" makes a dramatic pause, looks his distressed wife straight in the eyes, and says flatly—"No!" A large group of young men in the theater burst instantly into hoots, cheers, whistles, and applause. The young "godfather" has faced the crisis of feminine conscience and his manhood remained intact.

There are any number of ways we could analyze the meaning here. Marlon Brando, who played the part of the old "godfather," was reputed to have said that the film was itself a statement on modern industrial society. This idea is worth examining for its psychological and political implications. (And it certainly is worth noting that the young men who hooted and cheered were black; that is, gender identification may have more bonding power for males than does racial identification, even in terms of a movie with racist undertones.) One way to proceed is to look at what manliness means and has meant in terms of recent political figures in America.

Let's begin with Richard Nixon. From the memoir of Bob Haldeman, the story has come down that during the Vietnam War Nixon sent Kissinger to tell the North Vietnamese that he (Nixon) was *crazy* and would drop atomic bombs on Hanoi unless some important concessions were made.[I] Now if we were to imagine Kissinger as Harry Truman's national security advisor, and if we tried to picture Truman sending Kissinger to the North

I. For another account of Nixon's "insane flirtation with the actual use of nuclear weapons," see James Carroll's *House of War*, especially "The Madman Theory," pages 349–55.

Koreans or Chinese with a similar message, everything could be roughly the same—except that Truman would never have included anything about his being crazy. Truman could conceivably have made an atomic threat to the Koreans or Chinese during the Korean War (and maybe he did), but it is not probable that he would have cast doubts on his own sanity. Indeed not! Truman was the epitome of masculine self-assurance and control. He lived up to the image of a tough and decisive male leader, as his decision to drop atomic bombs on Nagasaki and Hiroshima reveals. To have questioned his own morality, judgment, or sanity was outside the realm of the acceptable. Truman was not a man given to nagging doubts or conscious uncertainties. One need only recall his aphorisms to see the point: "The buck stops here" and "If you can't stand the heat get out of the kitchen."

Dwight Eisenhower was in a sense a throwback to the benevolent paternalism of Franklin Roosevelt, though without Roosevelt's whiz-kid penchant for experiments in social engineering. Ike probably comes closest to the embodiment of the liberal, Western conception of God: a quiet golf-playing gentleman who allows the boys to pursue their business, but who will, in a pinch, step in to deliver a sermon or referee the game when the situation gets a little too unruly. John Kennedy was about half Harvard preppy and half Irish godfather. His dabbling in the classical arts as official patron indicates the former, and his handling of the Cuban missile affair points to the latter. This combination of traits—equally aristocratic at the opera and at war games—made him something of a charismatic figure. It should be remembered that it was John Kennedy who created the Green Berets, an elite military force specializing in stealth and subversion.

Lyndon Johnson was part Harry Truman and part messenger-boy-cum-manipulator. He was fond of earthy women and was not a man to wear his underpants too tightly. In choosing Hubert Humphrey for his running mate, Johnson made a very shrewd political move: Humphrey not only had his own brand of Midwestern charisma and a large constituency, he also was a man loyal to a father figure and therefore would not challenge Johnson's masculinity.[II] Historically, this meant that Humphrey would not resign the vice-presidency even though he was inwardly opposed to the Vietnam War. Humphrey thus lost credibility in the eyes of liberals and the young; in the process he lost his only opportunity to

II. See Chapter 19, "Orator of the Dawn," especially pages 458–59, in Robert Caro's biography of Lyndon Johnson, *Master of the Senate*, for a snapshot of the Johnson/Humphery relationship.

be president. Had Humphrey resigned in 1967, the Vietnam War could have been over by 1969—and Humphrey could well have become one of the most respected presidents of the century. In a sense, the Johnson-Humphrey administration represents a turning point in American politics. In a period of industrial depression, an experimenter like Franklin Roosevelt could operate on the great body of politics as an economic surgeon without appearing to be unmanly. But in periods of prosperity, those men who advocate broad-range social programs for the benefit of the unprotected, the left-out, the poor, and the environment, while neglecting to concentrate on industrial productivity and military might, are perceived as unmanly. And not only are they perceived as unmanly, but they begin to act "unmanly." That is, they begin to act indecisively.

It is for this reason—it was unmanly to come out against the Vietnam War—that Hubert Humphrey failed to resign the vice-presidency. He was unable to make the choice he knew he needed to make: the truly manly act, the truly *human* act, would have required a break from the military-industrial fraternity. Humphrey's failure to resign cost him the presidency while it cost the Vietnamese several more years of agony. Johnson, in the meantime, could fight the War on Poverty because he was also fighting the more macho war on the Viet Cong. Nixon, in turn, had a more ideologically masculine character than Johnson, although he was physically less manly. Latent Quaker scruples may have informed Nixon that nuclear war *is* madness; it is only with that possibility in mind that his instructions to Kissinger, if the whole account isn't apocryphal, make sense. But in the arena of economic policy, Nixon had no such Quaker reservations; he was a red-blooded American capitalist all the way. Nor should we neglect to notice that what many people call "the first September 11"—the military coup that overthrew Chilean President Salvador Allende in 1973—was promoted and assisted by the CIA during Nixon's presidency.

It's important to recognize an essential psychological undercurrent here. It is nearly impossible for many men to abandon the twin utopian projections of economic growth and military power without becoming confused, disoriented, and anxious. We often fail to understand to what extent the very identities of large numbers of men are tied directly into the economic-military-political system. This system is the eternal, objective sphere of masculine action and prerogative; without it, even presented with its theoretical contraction, many men feel threatened at the roots of their identity. The loosening of this ideology amounts to identity loss and identity crisis: hence the raw anger directed toward advocates of an ecological, no-growth, socialist economy.

So what about George McGovern, Eugene McCarthy, or George Wallace? Let's look at McCarthy first, briefly. McCarthy was a Catholic intellectual whose views on political matters were tinged with brilliant cynicism. He saw posturing and self-seeking in everyone, and his lofty morality was infused with sophisticated contempt. That's why he appealed (in a double sense) to the young: on the one hand, younger people appreciated McCarthy's caustic wit; on the other, there was a freshness in the young that provided fuel for McCarthy's vision and hope. Despite his inability to reach out to average American adults and speak to them in a language they could understand—his educated, ethical urbanity bounced right off the hardhats of the working class—McCarthy was perhaps the only man in recent political history whose demeanor, at least, fit the profile of statesmanship.

George McGovern offers us the prime example of the fate of a male politician who wanders off the beaten political track into the wilderness of identity diffusion. It's not as if McGovern didn't raise important issues. He did. McGovern was attempting to fight the battle against militarism from which Hubert Humphrey had retreated. Moreover, McGovern stands as the first major male candidate since the Second World War who dared to address on the highest level those issues relating to domestic politics. McGovern, more than any other Democrat since the Populists and Progressives, dared to diverge from the masculine military-industrial ethos into what might be called the feminine ethos of domestic well-being. And McGovern plainly embarrassed most men. His awkward and incompetent fumbling of the Eagleton affair finally provided the excuse for a wholesale abandonment of the McGovern candidacy. But most men turned from McGovern not so much because he made a political *faux pas,* but because he betrayed some deeply-seated and largely inarticulate masculine patterns of identity: McGovern was too interested in the "women's work" of domestic well-being and not interested enough in the tough "men's work" of industrial growth and military strength. Once again, we bump up against the resentments of the hard-hat working class, complete with flag decal and biblical convictions.

Perhaps, even more deeply, it was not McGovern's interests that embarrassed so many people but his implicit apologetic manner, for it seems to be extremely difficult for most men, at least in the United States, to advocate or openly support policies that run counter to those of the military-industrial fraternity without pronounced feelings of embarrassment and insecurity. Most men will hold back from the policies of the eco-feminist Left until either very strong male or very strong female politicians lead them there. But no politics has been, or will be, less rewarding than that of being a political wimp.

George Wallace can best be understood in comparison to Tom Watson, the man who was William Jennings Bryan's Populist running mate in the bizarre presidential election of 1896. Watson, as a leading Populist radical congressman from Georgia, had sponsored all manner of progressive and radical legislation: the eight-hour working day, control of the Pinkerton "detectives," free rural mail delivery, direct election of U.S. Senators, nationalization of the railroads, and so on. (All of which suggests, too, that manliness had a different psychocultural base in the nineteenth century—a *folk* base rooted in the remnants of village culture that had not yet been devoured by the utopian system.) But after the failed marriage of the Democratic and Populist parties in the '96 elections, Watson went into seclusion and emerged several years later a blistering demagogue. Prior to 1896, he had been a forceful and consistent advocate for racial harmony; after 1900, he was as rabid a racist as could be found in the country. In both instances, both before and after 1896, Watson considered himself a populist. But his later populism was saturated with racial hatred and worry over the purity of white womanhood. One can only surmise that something on Watson's emotional side was so damaged by the '96 debacle that he began to be psychologically deranged, or that his lust for political success led him to utopian conversion. Many of the racist and reactionary tendencies associated with "populism" to this day are responses to the same forces that led to Tom Watson's breakdown.

George Wallace was only a shadow of this latter Tom Watson. Wallace's "populism" was hardly more than the capitalist work ethic with a racist skin. His only dynamic elements were explicit race hatred and implicit masculine vigilance against socialist wimps. In terms of concrete political or economic proposals, Wallace was saying nothing that Nixon wasn't saying more intelligently—if Mutually Assured Destruction, the bombing of Cambodia, or a "plumbers" unit may be called intelligent. The forces of racial hatred suffered a major blow in this country when Wallace was shot and physically incapacitated.

Which brings us up to Jimmy Carter. (We will glide right over the unelected Gerald Ford who, as Lyndon Johnson supposedly said, played too much football without a helmet.) Shortly after Carter was inaugurated, E. F. Schumacher was in the United States on a lecture tour. Carter invited Schumacher to the White House for conversation. That Carter would even have wanted to speak with Schumacher indicates an important dimension of Carter's mind: concerns over rural culture, excessive energy consumption, pollution, meaningful work, etc. That Carter largely failed to address these issues during his term in office, that he appointed people to cabinet positions who were indifferent, if not hostile, to these concerns, and that he came increasingly, month by month, to embrace the *Realpolitik* of Mutually Assured Destruction—all this indicates how Carter's personal inclinations were suffocated and how he came steadily to serve the interests of the dominant ethos. Given his initial interest in the issues of "appropriate technology," and given the forces that were constantly driving him toward the deployment of standard military-industrial policies, it is of little wonder that Carter appeared indecisive and why even many concerned liberals couldn't stand to vote for him again.

Carter, like McGovern before him, could hear and sense and feel the urgency of addressing issues of pressing social policy that had to do with the passage of the Equal Rights Amendment, racial justice, equitable distribution of wealth, adequate health care, solar energy, organic agriculture, and the like. But those issues were on the agenda of social well-being and ecological health; they were not on the agenda of industrial expansion, national prowess, and utopian ideology. Carter, unlike McGovern, was a clever enough politician to get elected to the White House. But in the face of the real power centers in America, Carter's personal voice, the voice of spiritual honesty and social health, became muted. He emerged as his own ventriloquist. Thus, he said and did rash things simply to avoid political catatonia: the rescue attempt of the hostages in Iran, for instance, or the grain embargo imposed on the Soviet Union. For Carter to have addressed the real issues in a direct and forthright manner would have required a break from both the centers of masculine power and the traits of standard masculinity. Such concerted action would, no doubt, have brought him into the throes of an identity crisis. It would also have created a major political crisis, not only in this country, but worldwide.

This is an extremely important point. A truly progressive presidential candidate, whether successful in electoral terms or not—but especially if successful—would shake up the global power structure. Consider what would happen if the United States began to disarm; if democratic Third World revolutionaries were, if not actively supported, at least not opposed by American policy, weapons, and covert action; if the president stood strongly for reduced energy consumption, recycling, the phasing out of nuclear power, socialization of major corporations, abolition of the Federal Reserve system, reconstruction of small-scale parity agriculture, nonracist decentralization of schools, rebuilding of railroads, and so on.

All this would add up to a radically new political perspective, the kind of policy we would get by integrating the best of "Rainbow" and "Green." And the closest we have come to a male candidate who represents this perspective from within the "major" parties is, of course, Jesse Jackson—nor is it insignificant that Jackson comes from an ethnic minority that knows a long history of discrimination and oppression, and that in Jackson's youth he had a double springboard of spiritual training dovetailing with his association with Martin Luther King. (As Democratic Party candidates vie for national attention prior to the 2004 presidential election—I write this in late October of 2003—vegetarian Congressman Dennis Kucinich of Cleveland, Ohio, has begun to articulate a visionary program for the de-empiring of America. Apparently, the late Senator Paul Wellstone of Minnesota, had he not been killed in a plane crash a year ago, would also have been a presidential candidate with a unified message. But isn't it striking that the best and bravest of candidates have been black, Jewish, Lebanese (in Ralph Nader's case), and now a white vegetarian "boy" mayor? This should tell us implicitly how strong a grip the dominant system has on contemporary and conventional white consciousness. Apparently one needs to be racially or religiously in the minority, discriminated against or, by some quirk of fate, otherwise ethically energized in behalf of real moral principles, in order to boldly and persuasively confront the blatant excesses and even criminality of the dominant system. Privilege has a price.)

Carter lost in 1980 because people were sick of his Sunday-school demeanor. This reflects the inherent problem for liberals who are bound by the utopian ideology of capitalism in a world that can no longer tolerate such an ideology. Incapable of formulating an adequate noncapitalist and nonutopian structural analysis, liberals revert to moral clichés. But moral clichés (the energy crisis is the moral equivalent of war: Jimmy Carter; science must teach us the future: Walter Mondale) are no substitute for clear thinking and clear policy. One might even go so far as to say that liberal capitalism is little more than moral cliché. The candidacy of Walter Mondale confirms this perception. When the American people were asked to choose between an unequivocal capitalist like Reagan and a moral-cliché capitalist like Mondale, they overwhelmingly chose Reagan; he, at least, stood for something, even if it was theatrical imperialism and conventional greed. And if George Bush the Elder fascinated us with his muscular, slippery stealth, Bill Clinton simply hid, or tried to hide, his liberal indecisiveness under a massive and cunning policy theft from the Right. His truly famous charm was the lubricant that not only permitted his rightwing policies to glide effortlessly over his leftwing persona, it also enabled him to slip out of every box the Republicans set for him or in which they thought him trapped. Clinton was far less a stubborn mule than a greased pig.

But another serious question emerges at this point: whether the American people, so saturated with utopia, capitalist ideology, the fetish of consumerism, and so spiritually idolatrized, are ethically and politically *capable* of supporting the deeply radical provisions of an authentic Green and Rainbow policy. But in this quandary, too, one great truth stands out: unless this alternative is unapologetically put forth in political debate, the country as a whole will fall prey to an imperialist/fundamentalist form of fascism. (How long before we begin to hear that democracy is itself a "failed experiment" in Secular Humanism? That ecology is "pagan"?)[III]

America is a capitalist empire, the utopian machine uniquely unfettered by eutopian culture or spiritual restraint. (Eutopian culture did not have enough time to become deeply rooted in the American landscape before explicitly civilized structural formations, in accelerating industrial and technological modes, overpowered what eutopian culture there was. In addition, the spirituality of Christian America was overwhelmingly literalist, expansionist, and imperialist. In the face of these two forces—civilized power and religious arrogance—the formation of eutopian culture withered on the vine.) But, like all empires, this American one will also not last forever. The question looming on the horizon is not whether this empire is permanent, but in what way it will disintegrate. Will it fall through technological dysfunction, economic breakdown, ecological disaster, or military mayhem? Or will it become transformed, with the least possible

III. For an exploration of these very themes, see Chris Hedges' *American Fascists: The Christian Right and the War on America.*

suffering, through a deeply spiritual, ecological, socialist, and democratic consensus? Is there a eutopian politics that could "manage" the nervous breakdown of utopia with a minimum of carnage? There are at least four major tests by which to measure the probability of the latter alternative: whether feminism becomes both Rainbow and Green; whether Rainbow and Green can integrate and fuse; whether the Left can grow beyond its urban-industrial utopianism and utilize its powerful intellectual tools in the service of eutopia; whether working class men will be able to shrug off their macho posture and stand solidly within an ecological Rainbow strong with feminine values.

If this degree of change, this intensity of transformation, seems improbable and fanciful, then we must soberly consider its likely alternatives: the transformation of society in its entirety into something resembling the terrifying utopian sterility depicted by the late Stanley Kubrick in his film version of Anthony Burgess's novel *A Clockwork Orange*: human beings totally devoid of eutopian humanity; human beings reduced to replicas of empty utopian images; human beings with assertive, raw appetites, spiritually dead.

Civilization is not only the enemy of agriculture, it is also the destroyer of virtually all forms of culture incubated in the extraordinary long history of noncivilized social evolution. Civilization in its global hegemony devours and destroys all authentic culture. All human action, all human behavior, becomes subject to utopian mutation. Our lives become defined by institutional dependencies, gridworked in a vast interpenetrating saturation. At both the leading edge and power center of this utopian consolidation stands the hegemony of American corporate interests, protected by a colossal military, virtually unchallenged politically by any residual formation from the Left.

Congressman Jesse Jackson Jr., in "George Bush's Democrats," in the January 22, 2001, issue of *The Nation*, talks about the "conservative bipartisan coalition" in congress. He says it is this coalition that

> . . . allows Ralph Nader to say that we have one corporate party with two different names. If Democrats go down this bipartisan path it will only strengthen Nader and the Greens for 2002 and 2004. The move down that path has already been aided by Democrats: In 1992 a conservative Democrat, Bill Clinton, selected an even more conservative running mate, Al Gore, who in 2000 selected an even more conservative running mate, Joseph Lieberman. By helping

> to shift the Democratic Party and the country further right, a very conservative George W. Bush could select an ultraconservative Dick Cheney as his running mate—and win.[1]

So why this steady, even rapid, rightward momentum? There is a complex of reasons, a bundle and a tangle. Among them, oddly enough, is the collapse of the Soviet Union: for although the existence of the U.S.S.R. "justified" a militant capitalist antagonism, the disappearance of the U.S.S.R. paradoxically permitted militant capitalism free rein rather than (as naive common sense would have suggested) a welcome opportunity to relax. The surging globalization of capitalism, with the weakening of unions via deindustrialization, the reduction of agricultural commodity prices to (or below) Third World levels, the importation of massive amounts of sweatshop goods, the absolute supremacy of U.S. military capability as the enforcer of international economic and political inequality—all this (if we can achieve the detachment to contemplate the particulars in adequate perspective) is the global victory of utopian civility: the "mapping" of human genes, the space program (with its secret military aspects), the celebration of artificial intelligence, genetic engineering and robotics—a triumphant projection of fanatic immortality even as ecological and cultural disasters intensify and threaten to engulf the planet. In the center of this matrix is the alienated male, driven to achieve utopian immortality, brushing aside all hesitations and uncertainties as feminine weakness—wimps, hicks, hayseeds, and pussies who lack the drive, the balls, to be real gods. Virtually all the imagery of advertising and "cutting-edge" entertainment projects the power of virile perfectionism, of indestructibility—and, to prove the point, all that is imperfect and weak deservingly suffers the brutal consequences.

So, here, on the Earthbound side of reality, Green politics demands a totally banal reduced consumption, a far greater conservation of resources, hugely constrained pollution, and a far more localized economy. The dismissive contempt that informs large numbers of people in regard to the Green agenda is played against a psychological backdrop called "backwardness." That is to say, what Green politics requires in the direction of Green culture works up against an easily aroused popular refusal for a "going backward" rather than a "going forward." This is where "the vision thing" that so puzzled Bush the Elder comes of necessity into play. And although there can be—there is—a multilayered composition to the prevailing hesitancy regarding the Green agenda, one element is decisive and critical. While it may be possible to call this element by a variety of names, the term that seems immediately most useful is—gender.

Women want nothing more to do with washboards and an unending (or seemingly unending) series of babies. Women, week by month by year, are collectively moving to accrue political control over their destinies as female persons, which is, of course, what the endless fight over reproductive rights is about. For women, there will be *no going back*; therefore, Green culture waits on a sea change of relationship between women and men—that is, the dissolution of utopian male identity and the emergence of a far more equalitarian balance between women and men. Therefore, Green politics is largely reduced to an almost ineffectual carping at the fringes and margins of the prevailing system, precisely because women and men have not achieved the requisite reconciliation. If men are so stupid as to ride the utopian stallion into the omnivorous jaws of raging nature, women will watch with amused chagrin as their boys, full of self-absorbed male identity, choose disaster over transformation. We know that rain falls on the just and the unjust; so we can safely conclude that Global Warming (for instance) doesn't care in the least whether large numbers of women or men are outraged by the "lifestyle" implications of Green politics. But people, both male and female, who have acute environmental and ecological consciousness *do* care. And so "the vision thing" becomes more and more critical.

The dismal proportion of votes garnered by Ralph Nader in the 2000 presidential election can be ascribed to a number of factors. That is, it's probable that Nader would've received a substantially larger number of votes (enough to reach the magic five percent threshold, in regard to federal election funds) had Gore not been such a catastrophe as a supposedly "environmental" candidate. This is a contradiction inside a paradox: for Al Gore is supposedly a fine environmentalist who actually believes Earth's future lies in the balance; and one would think that Gore's lack of environmental passion would've caused a flood tide of citizens to vote instead for Nader. (We might even consider that possibility an inconvenient truth.) But, lo!, in the other camp rode the grim reaper on a "compassionate conservative" horse, saddlebags spilling cash and ads dripping fuzzy endearment. Therefore, a vote for Nader was a "wasted" vote, an act of political betrayal. Nader himself became a hated "spoiler," a smug brat, a scapegoat. That's one level.

Another level is Nader himself. The man is a political monk. His entire adult life has been a constant spiritual battle with Powers and Principalities. And so while Nader could unleash astonishingly brilliant condensations of complex political, economic, and environmental entanglements (with slogans to match, like "Corporate Welfare"), his very brilliance hid, but did not remove, the almost total absence of Green *culture* in his vision. The vision thing reaches beyond the sterility of both Republican and Democratic imaginations, beyond conventional Left and Right. More of everything—at least not less of anything—may not qualify for "the vision thing," but "prosperity" is a sacred buzzword that plays both to our trained appetites and to our induced fears. We have been programmed to love images of technological perfection, to crave possession of utopian toys. The felt risk in Green politics is doing without prosperity, without perfection. On the other hand, "Corporate Welfare" is an old-fashioned battle cry that can be followed by millions of people *only if* the leadership mushrooms accordingly. But we need a cultural vision to work toward, to be energized by: dry political formulae, no matter how abstractly true, do not substitute for the vital emotional content of cultural vision. This cultural vision cannot rely on eutopian fiction (like Ernest Callenbach's *Ecotopia*) as a prime mover; it must be undergirded by living examples of intentional alternative community that, while they may not have *solved* the problematic gender inheritance, are honestly struggling with it in healthy and creative ways.

Once again, we confront a paradox within a contradiction. If we postpone or avoid risk by acceding to corporate normality packaged in images of utopian perfection, we (a) enjoy a certain conventional prosperity for the immediate future, and (b) guarantee that the magnitude of future-wrenching will be horrifically greater. Therefore, there is a real urgency for those who recognize this need to get on with it in their daily lives—a ratcheting up of intellectual comprehension into lived experience. So the vision thing involves a complexity of factors that have not yet congealed. We know, for instance, that climate change is going to stuff SUVs down our throats. It's only a question of time—and not a whole lot of time, either. We know that women are reaching for at least political parity with men. We see it coming, we see it growing, but we don't know how long it will take for feminist politics to displace the confrontational masculine mind-sets and the predatory, perfectionist economic policies that "require" a massive military as a protection racket for the climate-change machine. Both of these developments are growing—environmental limitation and women's political power—and they may even be growing proportionately, if anyone is capable of making such an assessment. (Add into this mix global ethnic blending and the increasing dialogue between spiritual traditions.)

We *will* enter into a larger embrace of Green politics in the twenty-first century. We will be *forced* into it by ecological and environmental disasters. And we will be *guided* through these disasters by a political leadership that increasingly wears a female face—a face of the gardener, a face of the teacher and politician, a face of international consciousness and law—the caring face of a mother rather than the calculating face of a father. What will enable a Green humanity to avoid the gender rigidities of "backwardness" is the stubborn and determined socialism of sharing. "Backwardness" is, anyway, largely the prolonged cultural wreckage of the expropriated agrarian village, the folk-culture blowback of the aristocratic male superego. That is, the relative squalidness of the peasant village—insofar as peasant villages were squalid—must be seen in direct relation to both its oppression and its expropriation by utopian civility. Green *culture* will unfold exactly as women and men reconcile in a new ecological configuration governed by economic equality and environmental restraint. The precondition is the abandonment of macho posturing, the shrinking of the ordinary street version of the masculine superego—the hoots, jeers, and whistles.

In the March 2001 issue of *The Progressive*, John Nichols, in a review of *The Next Agenda: Blueprint for a New Progressive Movement*, edited by Robert Borosage and Roger Hickey, says that "In the midst of the most devastating agricultural depression since the 1930s, rural issues are—with the exception of a paragraph in a broader piece on economic issues by Jeff Faux—virtually ignored. The gap is dramatic and disturbing." This gap, I would add, is due precisely to the chronic, remorseless, and unrelenting belittlement of rural life and agrarian culture that simply saturates utopian civility. Is it any wonder then, that in the "course of three separate interviews with Gore during the 2000 campaign," Nichols

> . . . asked the Vice President about the anti-globalization protests, the farm crisis in the upper Midwest, the mass outcry over genetically modified foods, and a host of other fundamental issues. Gore refused, at every turn, to go near policy positions that posed any significant threat to Wall Street—even though such stances would

> have made his candidacy far more attractive to the nearly three million voters who supported Ralph Nader's Green Party challenge and the tens of millions of citizens who simply did not cast ballots.[2]

The vision thing threads its way through these dynamics. Green culture is imaginable only in proportion to the genuine abandonment of conventional male prerogative and the overcoming of addiction to the imagery of perfection. All else is postponement and make-believe. The fate of the Earth rests on the creation of Green culture, and the creation of Green culture waits on the shrinkage of the utopian male superego—even as global climate change bears down on us. Our dallying in the face of this disaster is both proof and measure of our addiction to civility.

NOTES

1. Jackson, "George," 14.
2. Nichols, "Next," 40.

18

A Baroque Apotheosis of Geopolitical Cretinism

I'M FOND OF OUTRAGEOUS titles. It is, perhaps, a fault. This one comes from a wonderfully apt turn of phrase, a marvelous compaction of well-deserved anathema, in Anatol Lievan's "Liberal Hawk Down" in the October 25, 2004, issue of *The Nation*. Thinking ahead to the election on November 2, I'm well aware how few people will join me in voting for either Ralph Nader or David Cobb. So it'll be either Bush or Kerry who wins. That's a given. My analysis suggests Bush will achieve a second term, or, at least, a start on a second term. But I am no electoral soothsayer. I have no crystal ball. There may be enough citizens disgusted with the current baroque apotheosis of neoconservative geopolitical cretinism that Kerry will be elected, if promising to win the unwinnable war on Iraq actually is the opposite of baroque cretinism.

It is my conclusion (and I confess to a certain historic recklessness) that a Bush victory might well result in an earlier and more complete discrediting of American Empire than a Bush defeat. In defeat, Bush returns, presumably to Texas, with a fat pension and a library, while the rabid Right goads Kerry into fulfilling the war promises that cannot be realized, thus ensuring an inevitable discrediting of the (obviously complicit) Democratic Party. In victory, however achieved, Bush has to carry his own gasoline into the houses he's determined to burn down, even as they're already enveloped in flames. If Democrats have enough ethical stamina and moral backbone, there is the possibility of impeachment and perhaps even conviction, thus disabling the reactionary, messianic Right for decades to come. A Kerry victory could be a disaster for the Democratic Party. A Bush victory virtually guarantees a Republican disaster.

I am far less interested in seeing Bush removed from office (though he is unfit to be there) than in seeing neoconservative hawks driven permanently from their perches of apocalyptic meddling and global catastrophic design. In order for this long-term discrediting to occur, the utter stupidity of neoconservative geopolitical cretinism has to sink in on the American people, pushed into their consciousness, rubbed in their faces, perhaps even over the Fox News Network. My biggest worry is that hope for a near-term contraction of messianic Empire is unrealistically optimistic. Yet, I believe with Lewis Mumford that the "traumatic institutions" integral to historic civilization must be transcended and discarded. Class and War, as Arnold Toynbee said, must be abolished. I believe such transcendence and abolition is possible and can happen, even if such unfolding seems the political equivalent of Slow Food.[1]

Transcendence and abolition will occur only as a result of profound spiritual conviction and ethical deepening in the direction of servanthood and stewardship. Politics alone will not get us where we desperately and urgently need to go. Are we ready to follow Baptist theologian Walter Rauschenbusch into the kingdom of God? Are we ready for ecological stewardship and socialist servanthood? How much suffering does it take to convict our hearts? How much disaster to crush our illusions?

We're in the throes of something big. The quickest, safest, and cleanest way through this mess comes by way of repentance. But repentance, as Malcomb X once said, is the hardest thing in the world, especially for those who are so absolutely certain of their political righteousness and spiritual justification.

1. This little essay was originally published on Wednesday, October 27, 2004, in a local shopper, here in northern Wisconsin, called the *Merrill Foto News*. The following Friday (October 29), an article in the *Wausau Daily Herald* pointed out that the children's news magazine *Weekly Reader* has been conducting polls of children, prior to presidential elections, since 1956, and the kids have never been wrong. They chose, of course, George W. Bush.

19

Domestic Stability Problems in the Alpha-Male Den

"Cartoons have an unfinished look that leaves a lot of interpretive space," writes Richard Goldstein in "Cartoon Wars." The graphic that illustrates his article in the February 21, 2005, issue of *The Nation* also adorns, in its cartoonish ugliness, the cover of the magazine, with the brazen headline "OFFENDED? Why Cartoons Get Under Our Skin." So there is square-headed yellow Sponge Bob, with his crossed blue eyes, wacky smile, buck teeth and red tie, hanging on a grainy brown cross. Sponge Bob on a cross!? Offended? Cartoonists, says Goldstein, "can sniff the emotional wind. The best of them delve into the subconscious." They create images that "work by tapping into hidden feelings."[1]

I invoke Richard Goldstein's language here, in this cartoonish context, because I want to sniff the emotional wind and maybe delve into the subconscious a wee bit with two recent books—George Lakoff's *Don't Think of an Elephant!* and Dana Priest's *The Mission*. To do this is to tread on spongy ground, and it invites harsh rebuke for, perhaps, inventing a sponge man whose "inner meaning" can conveniently be squeezed out. I am sensitive to the charge but, with all the risks of cartoonish simplicity, it seems to me the opposite is actually more the case. That is, we restrain ourselves from delving into the obvious because to do so is to undermine, or at least threaten to expose, the brutal, untenable conventions on which our daily patterns rest. We are afraid of eruption and disruption even as our daily patterns are larded with veins of violence and threats of violence. We keep our heads down for fear things will only get worse, that there will be hell to pay. This emotional doorway is lodged in the subconscious.

Let's go in. We'll start with a two-sentence quote, featuring Saudi ambassador Prince Bandar, from Craig Unger's *House of Bush, House of Saud*: "But during the Clinton era, Bandar had lost clout. Never an insider in the Clinton White House, he had disliked what he called the 'weak-dicked' foreign policy team of the Clinton administration."[2] "Weak-dicked" obviously means erectile dysfunction, which in turn means, in Bandar's lingo, that Clinton's foreign policy team was not manly. It was lacking in testosterone. It was weak, droopy, and flaccid. One term doesn't make a case for how strength is sexualized, or how the sexualization of strength undergirds power arrangements in our culture. So here are more such terms and allusions from Dana Priest's *The Mission: Waging War and Keeping Peace with America's Military*, which is largely on how the military has become the real foreign policy arm of the United States. We learn that Donald Rumsfeld and Colin Powell were seen as "adults," "father figures with iron fists." Rumsfeld, especially, was gruff, autocratic and impatient, projecting an "often patronizing, father-knows-best way" of talking to subordinates.[3] We learn that Bill Clinton, a pot-smoking draft dodger, was resented by the military, and how, when president, thirty years after his evasive antiwar behavior, "he seemed cowed by his past, afraid to really challenge the military on its shape or priorities for fear it would revolt publicly."[4] *Cowed*?

Priest quotes General John Shalikashvili: "You know—real men don't support Clinton." (The same general says "Real men don't do MOOTW"—Military Operations Other Than War. Perhaps MOOTW is the sound cowed antiwar politicians make as they try to herd lightly-armed soldiers into peacekeeping.) Marine General Anthony Zinni is quoted as saying, "If it doesn't draw blood, it ain't a sport." We are told that Zinni liked "Arab culture; it reminded him of his Italian heritage." He developed "great respect for the Saudis and their monarchy." "The treatment of women and the presence of imported slaves were taboo subjects." Later, we see Zinni escorted to the Koi-Tash Army Base to watch a "peacekeeping" exercise, where "a team of U.S. Army Special Forces troops had spent four months training their Kyrgyz counterparts in 'peacekeeping' methods, a euphemism for lethal tactics." This exercise was followed by the watching of a video that "showed a female singer, clad in a leopard-skin bikini, singing the Kazakh national epic poem in sacred tribute to Kazakh soldiers."[5]

On pages 49, 50, 75, 79, and 103, we can gather, from scattered passages, how the "blights" of drug trafficking, criminal networks, and terrorism developed in countries with enormous income disparities, that all these problems transcended national borders and flourished in unstable, unjust political systems. And while the Pentagon assisted armies fighting drug traffickers, taught counterinsurgency techniques in countries with—this phrase simply must be quoted—"domestic stability problems," no one ever talked, for instance, about leaning on Middle East monarchs to allow democratic political change. Besides, "How could the leaders of these countries take U.S. pressure toward democracy seriously if most of Washington's handouts were for surveillance, weapons, and counter-terrorism training?"[6]

The second section in Dana Priest's book is called, simply, "The Special Forces." We'll indulge in a few more quotes, intersecting governmental policy with sexual identity, before moving on to George Lakoff's *Elephant!*, which we'll try not to think of in the meantime. John Kennedy, the great Camelot charmer, promoted the Special Forces. Here's Ms. Priest, a touch shaky on her date:

> In the mid-1960s, as President John Kennedy watched the popular challenges to postcolonialism from Third World rebels and leftist political parties, he came to believe they were all interconnected. He saw a worldwide insurgency sponsored by communist China and the Soviet Union, and he turned to the special forces to stop it. According to SF historians, JFK's appreciation for the independent mindset and skills of the special forces dated back to his wartime experience as a PT boat commander. He liked their style, too, and lifted the Army's ban on the SF signature, the green beret, modeled on the headgear worn by the British Royal Marine Commandos, so they could be easily distinguished from the regular army crowd.
>
> Kennedy's support, though, was much more concrete than endorsing a beret. In 1961, he dramatically expanded special operations, including Navy and Air Force units, into a worldwide counterinsurgency force. They would rely on the doctrine and tactics of unconventional war—jungle combat, PSYOPS, long-range reconnaissance, sabotage—to defeat guerrilla forces on their own territory.[7]

(Field Manual 31–20, "Doctrine for Special Forces Operation," issued in 1990, calls for Special Forces to "organize, train, advise, and assist" foreign militaries so they can "free and protect [their societies] from subversion, lawlessness, and insurgency.") The Special Forces team ("There are no women on Special Forces A-teams") recreates "the family, the tribe, the brotherhood." One Special Forces sergeant told Priest the biggest fear of any team member was "the possibility he might let down 'dad,' the team sergeant."[8]

> There are no women anywhere. But sex, or at least the talk of it, was a constant—in wall art, in conversation, and in jokes. Even

in otherwise boring Central Command's Special Operations Command briefings: "Special forces!" barked one briefer, his voice rising with enthusiasm. "We penetrate deeper, stay longer, and deliver a bigger payload!"

> The Alpha-male den in Kosovo is a relentlessly competitive place, with unforgiving physical standards and unspoken psychological codes of conduct. Most team members are conservative in outlook and disdainful of rank and hierarchy for its own sake. The job attracts risk-takers and daredevils, but commanders try hard to weed out "cowboys" who could endanger the team. It's similar to the esprit de corps found in the American firehouse.[9]

A newcomer to a team in Kosovo was routinely humiliated by the team sergeant who "used the female pronoun and spoke to him through a third party: 'Tell her . . . .'"[10]

Let's do two more passages here, before moving on to George Lakoff. The first is from Dana Priest's *The Mission*.

> "It takes something special to be in SF, you know," one pale, dark-eyed staff sergeant whispered at a restaurant in Kosovo one evening. "Not everyone wants to kill somebody. Not everyone could, you know. Not everyone has it in them. I do."[11]

The second is from a Patricia Williams' column, "Power and the Word," in the February 28, 2005, issue of *The Nation*.

> Consider that panel in San Diego at which Lieut. Gen. James Mattis, commanding general of the Marine Corps Combat Development Command, opined, "Actually, it's a lot of fun to fight. You know, it's a hell of a hoot . . . . It's brawling . . . . You go into Afghanistan. You got guys who slap women around for five years because they didn't wear a veil . . . you know guys like that ain't got no manhood left anyway. So it's a hell of a lot of fun to shoot them." Mattis, who is "responsible for developing Marine warfighting doctrine, techniques and tactics," is a man with a soft, puppyish, even endearing face very much at visual odds with the brutality of the statement.[12]

George Lakoff is a Berkeley linguist whose political analysis revolves around a concept he calls "framing." What are frames? They are "mental structures that shape the way we see the world." They are "part of what cognitive scientists call the 'cognitive unconscious'—structures in our brains that we cannot consciously access, but know by their consequences."[13] This language of the "cognitive unconscious" is about as close as Lakoff gets to psychology. He identifies himself as liberal and progressive, and his intent is to explain how the Right has, through gobs of money, think tanks and concentrated framing, captured political power in this country, and how progressive forces need to roughly replicate the Right's process in order to reframe public policy and begin to win elections. Right and Left, Lakoff says, have distinctive and defining parental models—"frames"—through which their respective visions are projected and received. For the Right it is The Strict Father. For the Left it is The Nurturant Parent.

What's most interesting and even crucial about Lakoff's analysis (though he waits until Chapter 7, "What the Right Wants," to tell us about it) is not only that the "right wing is attempting to impose a strict father ideology on America and, ultimately, the rest of the world," but that "God is the original strict father":

> Many conservatives start with a view of God that makes conservative ideology seem both natural and good. God is all good and all powerful, at the top of a natural hierarchy in which morality is linked with power. God wants good people to be in charge. Virtue is to be rewarded—with power. God therefore wants a hierarchical society in which there are moral authorities who should be obeyed.
>
> God makes laws—commandments—defining right and wrong. One must have discipline to follow God's commandments. God is punitive: He punishes those who do not follow his commandments, and rewards those who do. Following God's laws takes discipline. Those who are disciplined enough to be moral are disciplined enough to become prosperous and powerful.
>
> God is the original strict father.[14]

In Chapter 9, in a question-and-answer format ("Is religion inherently conservative? Are progressive ideals inconsistent with religious beliefs?"), Lakoff says that "One of the problems is that the progressive religious community, particularly progressive Christianity, doesn't really know how to express its own theology in a way that makes its politics clear, whereas conservative Christians do know the direct link between their theology and their politics. Conservative Christianity is a strict father religion." Lakoff goes on to tell us how the "strict father view of the world is mapped onto conservative Christianity":

> First, God is understood as punitive—that is, if you sin you are going to go to hell, and if you don't sin you are going to be rewarded. But since people tend to sin at one point or another in their lives, how is it possible for them to get to heaven? The answer in conservative Christianity is Christ. What Jesus does is offer them a chance to get to heaven. The idea is this: Christ suffered on the cross so much that he built up moral credit sufficient for all people, forever. He then offered a chance to get to heaven—that is, redemption—on the following terms, strict father terms: If you accept Jesus as your savior, that is, as your moral authority, and agree to follow the moral authority of your minister and your church, then you can get to heaven. But that is going to require discipline. You need to be disciplined enough to follow the rules, and if you don't, then you are going to go to hell. So Jesus, with his moral credit that he gained from suffering, can pay off your debts—that is, your sins—and allow you to get into heaven, but only if you toe the line.

And then Lakoff says that "Liberal Christianity is very, very different":

> Liberal Christianity sees God as essentially beneficent, as wanting to help people. The central idea in liberal Christianity is grace, where grace is understood as a kind of metaphorical nurturance. In liberal Christianity, you can't earn grace—you are given grace unconditionally by God unconditionally. But you have to accept grace, you have to be near God to get his grace, you can be filled with grace, you can be healed by grace, and you are made into a moral person through God's grace.
>
> In other words, grace is metaphorical nurturance. That is, just as nurturance feeds you, heals you, takes care of you, just as a nurturant parent teaches you to be nurturant and allows you to be a moral being, just as you can't get nurturance unless you are close to your parents, just as you must accept nurturance in order to get it, so all of these things about nurturance are true of grace in liberal Christianity. Nurturance comes with unconditional love, in the case of grace, the unconditional love of God. What makes a religion nurturant is that it metaphorically views God as a nurturant parent. In a nurturant form of religion, your spiritual experience has to do with your connection to other people and the world, and your spiritual practice has to do with your service to other people and to your community. This is why nurturant Christians are progressives; because they have a nurturant morality, just as progressives have.

> But at present nurturant Christians, Jews, Muslims, and Buddhists, in this country are not organized. They are not seen as a single movement, a progressive religious movement. Worse, secular progressives do not see those with a nurturant form of religion as natural members of the same political movement. Not only do spiritual progressives need to unite with each other, they need to unite with secular progressives, who share the same moral system and political objectives.[15]

As much as I am convinced that George Lakoff's "framing" analysis is, and will be, extremely helpful in guiding the Left's public policy articulation in the years ahead, there is a huge conceptual gap between (on the one hand) the "thirty years, billions of dollars, and forty-three institutes" that conservatives have used to "reframe public debate" and (on the other hand) God as the "original strict father" in a religious context going back thousands of years. Add into this the virtually total absence, in Lakoff's book, of how civilization since its inception has been governed by hierarchical, strict father morality, and how civilization's utopian obsession with transcendence (rising above and surpassing nature) has been threaded through economic ideologies and religious mythologies the world over.

That is to say, it's not that Lakoff is wrong about framing and reframing (I believe he's on to something profoundly important and very useful); it's just that his contextual embeddedness is alarmingly shallow. There is revolution afoot here, historic phase shift at work. Thirty years of money and a few dozen think tanks didn't create this dynamic. The modern "democratic" Right is the residual superego of traditional civility aided and abetted, supported and pushed, by the righteous superego of conventional religion under the influence of the monarchical model of God. So when Lakoff says that the progressive religious community "doesn't really know how to express its own theology in a way that makes its politics clear, whereas conservative Christians do know," we have to realize the enormous burden of reformulation, particularly in the area of political ethics, that "progressives" are faced with. In terms of the Christian Gospels, this is nothing less than the rediscovery of the kingdom of God as the core teaching of Jesus.[I]

I. On pages 59 and 60 of *The Story of Utopias*, Lewis Mumford says the "utopia of the first fifteen hundred years after Christ is transplanted to the sky, and called the Kingdom of Heaven. It is distinctly a utopia of escape . . . . So the utopia of Christianity is fixed and settled: one can enter into the Kingdom of Heaven if a passport has been granted, but one can do nothing to create or mold this heaven. Change and struggle and ambi-

Lakoff can assert, at the end of Chapter 9, that "Progressive activists—for living wages, women's rights, human rights, the environment, health, voter registration, and so on—are American patriots, working with unselfish dedication at making a better world, a world that fits fundamental American values and principles," and he can also assert that

> Progressive thought is as American as apple pie. Progressives want political equality, good public schools, healthy children, care for the aged, police protection, family farms, air you can breathe, water you can drink, fish in our streams, forests you can hike in, songbirds and frogs, livable cities, ethical businesses, journalists who tell the truth, music and dance, poetry and art, and jobs that pay a living wage to everyone who works.[16]

And yet this patriotic apple pie—if we actually baked and ate it—is tantamount to a eutopian political revolution.

There are enormous implications here. First, to grasp Lakoff's prescription in this larger context is to see that the current congealing, on the Right, of traditional civilization with traditional religion—both "strict father" models—results in a kind of heedless technological utopia bursting with fantasies of End Times rapture. The background implication is often overlooked. The steadily rising sea level of earthly eutopian values—mostly unrecognized, largely lacking in the capacity "to express its own theology

tion and amelioration belong to the wicked world, and bring no final satisfaction." It is precisely this impulse, so deeply embedded in Christian theology, that underlies the paralysis and passivity of even "progressive" Christianity. Overcoming this spiritual immobilization depends on the discovery of powerful earthly content in the kingdom of God as propounded by Jesus, a discovery that brings with it a radical reformulation of what's called God.

In section 6 of Chapter Six ("The Polytechnic Tradition") in *The Pentagon of Power*, Mumford says, on page 158, that in confronting the "quasi-religious cult of mechanization," polytechnics was "helpless" and lacked a "corresponding ideology to draw on." Why? "Since all the scattered trades and crafts and vocations had grown up over the ages, their underlying inner unity was largely an unconscious traditional heritage, and their values had not yet been translated into a philosophy—much less a common systematic method."

Put these dynamics together—the utopia of a Sky God and a quasi-religious cult of mechanization that derives from the Sky God's cosmic blueprint—and you have a megamachine on Earth, an ideology so pervasive that both the traditional heritage of polytechnics and the contemporary progressive religious community have no corresponding ideology to draw upon, and no way to express their own theology in ways that make their politics clear. In other words, the mentality of utopian civility has become ubiquitous and totalitarian. Utopia is a compulsive blabbermouth while eutopia is inarticulate to the point of being mute.

in a way that makes its politics clear"—is nevertheless felt intuitively by the Right. This intuitive alarm (father values *are* being threatened) results in an increasingly militant resolve to combat the elusive, underlying threat, often identified as "secular," "humanist," or "pagan." Political victory for the eutopian Green Left implies the permanent shrinkage of all strict father models in religion, business, education, and government, the termination of the Alpha-male den as the nucleus of power and assault, and the obsolescence of a punitive God on "top of a natural hierarchy in which morality is linked with power"—God as superego from which we need deliverance.

Lakoff is very much right about "framing" and "reframing." But his little thirty-year think-tank bubble is almost laughable in its analytic superficiality. (Compare it to Mumford's depth analysis in *The Myth of the Machine* and *The Pentagon of Power*!) And yet, despite its truly embarrassing deficiencies in history and psychology, George Lakoff's little *Elephant!* book is providing a key concept at a critical point in time. Let's just hope the Democratic Party doesn't smother this brilliant clarification in the worthless rags of political timidity and cynical calculation.

NOTES

1. Goldstein, "Cartoon," 7.
2. Unger, *House*, 219.
3. Priest, *Mission*, 25.
4. Priest, *Mission*, 43.
5. Priest, *Mission*, 44, 56, 69, 87, 115–16.
6. Priest, *Mission*, 103.
7. Priest, *Mission*, 135.
8. Priest, *Mission*, 129, 122, 125.
9. Priest, *Mission*, 130.
10. Priest, *Mission*, 131.
11. Priest, *Mission*, 123.
12. Williams, "Power," 10.
13. Lakoff, *Don't*, xv.
14. Lakoff, *Don't*, 81–82.
15. Lakoff, *Don't*, 102–3.
16. Lakoff, *Don't*, 110.

20

Carried Away by Joy

It's interesting and perhaps a little puzzling how writers like Edward Abbey and Richard Manning get their analyses right in regard to how the local and small-scale get hammered by the centralized and big (I'm thinking here, especially, of Abbey's essay "Thus I Reply to Rene Dubos" in *Down the River*, and Manning's book *Against the Grain: How Agriculture Hijacked Civilization*), and then totally confound their grounded insights by invoking "civilization" in a way that renders their insights and analyses powerless or at least intellectually bewildering, confounding, and compromised.

James Howard Kunstler, in his 2005 doomsday book *The Long Emergency: Surviving the Converging Catastrophes of the Twenty-First Century*, dabbles with the same confounded powerlessness as he lays out, in rather exhausting detail, how global oil depletion will squeeze wanton, reckless, mindless affluence out of mass human behavior and thrust us all (those who survive in our overpopulated world) into a space more like that of preindustrial society, except for the degraded landscape, the toxins, the depletions, the broken expectations, and the serious deficiencies in subsistence knowledge. Does Kunstler imply that we're headed back to traditional patterns of gender? I think the only reference to women's issues in *The Long Emergency* is this one: "Reestablished traditional divisions of labor may undo many of the putative victories of the feminist revolution."[1] That remark (with its ambiguous "putative") is about as explicit as Kunstler gets in regard to a feminist future. *The Long Emergency* is largely about what the world will look like, what we are likely to undergo in the remainder of the twenty-first century, as the inevitable oil contraction sets in, deepens, and intensifies—less a Dark Age than a Dim.

I have no basic disagreement with Kunstler's projection of the energy situation. (Though I am nowhere as neutral as he appears in regard to nuclear power.) In Kunstler's view, as far as the United States goes, the arid Southwest will lead in suffering, followed by the Southeast: the former's aridity is insurmountable, while the latter's quasi-tropical heat compounded by fanatic religious individualism will, possibly, result in the restoration of neofeudal plantationism. The Plains states to the Rockies will radically shrink in population, and the Pacific Northwest will be heavily burdened by population influx (from California, especially) and harassed by Asian pirates. The only part of the country somewhat capable of landing on its feet is the Northeast, from Maine to Minnesota, for reasons Kunstler spells out in his last chapter, "Living in the Long Emergency," pages 289–293. (Kunstler lives in upstate New York, so we can, if we wish, take his regional priorities with a grain of preferential salt.)

Here and there, always with positive intent, Kunstler uses words like "civilized," "civilizational," and "civilization." As a slash and burn ethnologist of suburbia (almost everything that's wrong with oil-glut America is summed up in suburban sprawl, "part of the greatest misallocation of resources in world history"[2]), Kunstler can't help but have read Lewis Mumford, whose comprehensive study of cities, of the origins of civilization, is too basic to have been missed or overlooked. But I find no ambivalence, irony, or ambiguity in Kunstler's conventional use of the *civitas* words. I find no slipping back along the trajectory of civilization, no hint that the debacle we face is the *consequence* of "civilized values." Civilization, for Kunstler, apparently remains the beacon, the phosphorescent miasma in the swamp, that will lead us to higher ground, even as we enter the political wilderness of oil depletion.

Kunstler misses two things. Or, rather, he misses one thing and fails to recognize the importance of a second thing. What he misses is Mumford's critical insight that the industrialization of society did not begin with steam engines in eighteenth-century England but, rather, with the rise of kingship and civilization over five thousand years ago. *Human beings* constituted the first machines, industrially organized, that performed the centrally-dictated work of the earliest civilizations. Their social mode was slavery. The modern "industrial revolution," in this understanding, is only the transferal of slavery to externalized machines. The "industrial revolution" did not invent regimentation; it only enabled regimentation to saturate and transform cultural forms into explicitly utopian rationality. The "industrial revolution" *externalized* slavery and, with its concomitant destruction of noncivilized cultural forms, extruded civilized organization as an imposed global norm.

What Kunstler fails to recognize is that globalization represents utopian civility in planetary form, so it is therefore possible to say that this crisis is not just another crisis in a long series of crises but something of an epochal culmination. So when he says he wants to preserve civility and civilization (he nowhere even vaguely defines these terms or invokes their history, just seems to presume the reader understands them as semantic flags indicating the high cultural ground in an impending flood of barbarism), he is implicitly acknowledging that either he fails to recognize this crisis as the product of civilization or, recognizing the causality, cannot bring himself to express it, either because he can't bear the realization or because he fears rude dismissal as an irresponsible crank. At the bottom of this failure lies the magical sanctity that civilization continues to carry and thrive on. It is a magical beast whose very victims crave to revive it when it appears to be failing. It is precisely this (what shall we call it?) metaphysical enchantment, this religious veneration, that is so unbelievably hard to break out of. Engulfed in utopia, we can see the wreckage, but we hurriedly attribute the disasters to recent flaws in design (i.e., we did not foresee the consequences of the "industrial revolution") even as Mumford teaches us (as, incidentally, Arnold Toynbee also teaches us) that the flaw lies in the original; that is, the flaw lies in civilization itself, in the structure of its origin. (Toynbee says "diseases." Mumford says "traumatic institutions.")

The failure to recognize the destructive attributes of utopian civility is partly explained by civility's magnificent, awe-inspiring, colossal creations. The "product" overwhelmed and obscured the price. But that's only part of the explanation and, at that, the more obvious and self-explanatory part. The deeper aspects of our failure to recognize the terminal consequences of civilization are much more troubling; they involve our semiconscious fears of the past, of backwardness, of darkness, of the primitive, of the countryside, of hard work, of dirt, and of death. Civilization supposedly lifts us out of and beyond all this. It rescues us from these negativities and is our utopian savior. It postpones death and, with its state religions, offers flight to heavenly bliss after death for those with right behavior and correct belief. Civilization is the conceptual package in which all these presents are to be unwrapped and exulted over.

It's not incidental that Kunstler's sixth chapter, "Running on Fumes: The Hallucinated Economy," is not only the most passionate and heartfelt, but that its theme is the "entropic mess that our economy has become" in the "final blowoff of late oil-based industrialism," this "new economy" that promised to "deliver heaven on earth, where everyone everywhere would be rich."[3] The third section of that chapter ("The High-Entropy Economy") has some of the finest writing in the entire book. Explicitly addressing entropy and how industrial economies simplify and even wreck natural systems in order to extract wealth, Kunstler says "Efficiency is the straightest path to hell."[4] But such phrases—heaven on earth, path to hell—are apparently only seen by Kunstler as colorfully expressive language, not for what the civilized, utopian economy actually imagines (heaven) and eventually achieves (hell). Civilization is at its core a false spirituality. True spirituality can be identified in the Four Noble Truths and Eightfold Path of elemental Buddhism, or in the kingdom of God teachings in the first three Christian Gospels. But, in the West, post-Constantinian salvationism became the primary tendency of orthodox state religion, so entangled with utopian, aristocratic, and civilized mythology that it is for all practical purposes inseparable from civilized ideology: inseparable, that is, unless one undergoes "conversion" to a "demythologized" spirituality rich in eutopian ethics. Such conversion moves one from fear to trust, from selfishness to sharing, from endless affluence to existential simplicity, from otherworldly obsession to this-world stewardship, from infantile clinging to utopian apparition to a weaned consciousness capable of eutopian community.

In some ways I like Kunstler's no-nonsense fact-laden analysis. (He says lots of useful things about gardens, small farms, local food production, the revival of rural life, and the survival of small towns.) But I dearly miss the deeper history, the more complex analysis. In Kunstler's view, the plug is about to be pulled in the American ant colony, and he's simply predicting (like Isaac Asimov's Hari Seldon?) how the ants will behave as a consequence. Here's a fair sampling of Kunstler's understanding:

> The circumstances of the Long Emergency will be the opposite of what we currently experience. There will be hunger instead of plenty, cold where there was once warmth, effort where there was once leisure, sickness where there was health, and violence where there was peace. We will have to adjust our attitudes, values, and ideas to accommodate these new circumstances and we may not recognize the people we will soon become or the people we once were. In a world where sheer survival dominates all other concerns, a tragic view of life is apt to reassert itself. This is another way of saying that we will become keenly aware of the limitations

> of human nature in general and its relation to ubiquitous mortality in particular. Life will get much more real. The dilettantish luxury of relativism will be forgotten in the boneyards of the future. Irony, hipness, cutting-edge coolness will seem either quaint or utterly inexplicable to people struggling to produce enough food to get through the winter. In the Long Emergency, nobody will get anything for nothing.
>
> I believe these hardships will prompt a return to religious practice in all regions of America, with tendencies toward extremism that will be worse in some places than in others. In the absence of legitimate or effective secular authority, church authority may take its place, perhaps for a long time to come. People desperate for legitimate authority to assist them in organizing their survival will probably accept more starkly hierarchical social relations in general and disdain democracy as a waste of effort. They will be easily led and easily pushed around. This, along with the emergence of a substantial agricultural laboring class, suggests that the ranks of society will be much more distinct in the Long Emergency, with far less movement between the ranks. Do not expect more social equality—expect much less.[5]

On the one hand, I think we have to acknowledge that Kunstler's dreary vision lies within the range of dreadful possibility. He implies that males everywhere will openly dominate (white males especially, except for the arid Southwest where white male supremacy will be challenged, perhaps successfully, by brown male supremacy), and women will find themselves trapped in "traditional divisions of labor." This also implies that gains in women's freedom, as with the relatively new freedoms for racial minorities, are hardly more than temporary cultural loosenings made possible by a relaxation in economic roles due to energy extravagance and industrial affluence: take away the surfeit and old patterns immediately (or almost immediately) will appear. You may have thought you've come a long way, baby, but you'd better think again. The same apparently is true for our experiment in democracy.

I believe, I dearly hope, that James Howard Kunstler is really wrong. Not that I think he's wrong about the energy funnel we're entering, big end first. Not that he's wrong about the fundamental importance of localized food production or the potential problems involved in heating our homes. There are lots of facts, observations, and assertions in *The Long Emergency* that are invigorating, full of brisk fresh air, and tough enough to cut right through much conventional feel-good crap. But there is no yeasty spirituality here. Kunstler's prophecy is devoid of such vision and conviction. The paradox I feel is that Kunstler himself would welcome ecotopia, eutopia, or (whatever else we may wish to call a truly ethical and ecological culture)—the kingdom of God. That is, I sense that in many ways Kunstler is personally already there; but as a professional writer, as a "public intellectual," he's apparently afraid to externalize, in writing, his internal conviction and condition. A richer view is not without historical underpinning.

The last two chapters in Mumford's *The Myth of the Machine* are rich in explication of what Mumford calls "democratic technics," including that (mostly unknown and unrecognized) contribution to waterpower and windpower developed in Benedictine monasteries in the early Middle Ages. And although Mumford is also clear that the "megamachine" that arose in Europe after the sixteenth century, with its authoritarian technics that steadily undercut democratic technics, has achieved totalitarian dimensions in our own time, there is very much the possibility that the accelerated spread of democratic conviction will counter and overcome the hierarchical inclinations that Kunstler gloomily forecasts. The future is not a done deal. As late as the sixteenth century, Mumford says, "more than half the recorded days were holidays; while for Europe as a whole, the total number of holidays, including Sunday, came to 189, a number even greater than those enjoyed by Imperial Rome."[6] I don't wish to underestimate the difficulty of our adjusting to radically more frugal conditions, including our serious cultural amnesia in practical democratic technics. But if one is not entitled to airhead optimism, neither is one entitled to a kind of ghastly fundamentalist predictability based on thrusting our meanest tendencies forward as cookie cutters for exhausted distress.

Here's the thing: we need the brutal honesty of Kunstler's energy analysis. I believe it's fundamentally accurate and, largely, prophetically apt. But we need to beware of how brutally honest analysis can be distorted as social and cultural projection when focused through the lens of an inner condition that evades and avoids, has no truck with, the realm of the spiritual or that deeper, more complex probing that Mumford is surpassingly good at. In the dreary projection of Kunstler's imagination, it is, apparently, only mundane *Realpolitik* that matters. It's only possible to extrapolate from the empirically verifiable. More subtle and elusive dynamics may be momentarily distracting, privately appealing, but to be credible one must stick to the most obvious of facts.

To be blunt, I don't like Kunstler's lack of historical depth, either in the manner of Mumford or of Norman O. Brown, and I don't trust his future projections, beyond a certain obviousness, especially when those projections are inadequately fueled or informed by historical, psychoanalytical, or spiritual profundity. Kunstler's future, fully granting the energy dynamics and social trends, is lacking in the unexpected. I, for one, anticipate surprise. I expect, amid all the suffering, to be carried away by joy.

NOTES

1. Kunstler, *Long*, 304.
2. Kunstler, *Long*, 233.
3. Kunstler, *Long*, 185.
4. Kunstler, *Long*, 191.
5. Kunstler, *Long*, 303.
6. Mumford, *Myth*, 271.

21

The Deferment Pit

WHAT WE SEE—WHAT WE don't see—at the lower, opaque level of consciousness is the basic human capacity to instantly and immediately detect hesitation, doubt, uncertainty, vacillation, or wavering. This uncertainty, compounded by inherited deference to established authority, is the bane of the liberal Left. The Left is a hat-tipping peasantry to the haughty aristocratic Right. Intuitively, all of us not only recognize this dynamic, even if dimly and reluctantly, we are also fully caught up in it politically.

The lack of moral authority on the Left can be said to be, in one respect, the Left's defining weakness. Where the Right can invoke God, the nation, the flag, the cross and patriotic obedience, the Left can only stammer about equality, equity, ecological integrity, and universal humanitarianism. We know this is true, and yet we have only the most paltry and threadbare explanations for this imbalance. Why, when the Left has virtually the entire heritage of the world's most profound ethical teachings by which to shape and inform its programmatic vision, is the Left so weak, so anemic, so devoid of moral force, so *apologetic* and deferential? The paradox, the contradiction, is deeply puzzling. There are, at least, four strands of analysis here: male entitlement/female deference; formal religion with its overwhelmingly divine male imagery and fierce emphasis on personal salvation; secular governance with its ancient male presumption and warrior predominance; economic aggression with men depicted as wresting wealth from stubborn nature. These strands constitute many of the fibers that weave the status quo and prevent the ethical vision of the Left—servanthood and stewardship, in broad outline—from becoming a tougher fabric than gossamer idealism.

Where the Right is in control of the emblems of authority, order, divine righteousness and wrath, all of it embedded in elaborate—and even, at times, elegant—sentimentality, the Left advocates the relaxation of authority, the dispersal of power, the trusting in organic order, and the understanding of divinity as inherently compassionate and self-giving. The Right fiercely guards its fearful selfishness in a cloak of militant traditionalism. The Left can only advocate for a world of sharing and stewardship by a difficult blend of spiritual vision and ethical realism—which must then cope with proliferating accusations of soft-headedness and (the oddest insult) "utopianism." Sentimentality is the underbelly of brutality. As a generality it is true that the Right's accusation of softheadedness against the Left is a deflected projection of the Right's fiercely guarded sentimentality. The Right's conventional accusation against the Left is an inversion based on denial; that is, the Right is not nearly so much hardheaded as it is hardhearted, and its claim to being hardheaded is fitting only insofar as it wears an ideological helmet: hardheaded in a way directly analogous to a football player's headgear.

What we can call the feminization of the Left has, in recent decades, served to immensely strengthen the complexity of Left analysis and to discredit, at least in the eyes (and votes) of many men, the Left in electoral politics. The essential *humanization* of the Left, by its inclusion of feminist analysis, minority critiques and ecological considerations, has not only made Left analysis more deeply comprehensive, it has also served, at least in the immediate short run, to provide precisely those hot-button issues that cause a large number of men, with overwhelmingly conventional male identity, to spurn and repudiate Left politics. That is, as Left analysis has become hugely more inclusive, it has simultaneously caused, by virtue of its very inclusiveness, large numbers of individualistic men to spurn their class identity in favor of a (admittedly elusive) political male identity organized against their class interests.[1] This is precisely what AM talk radio is designed to do. The Republican Right is more than happy to provide a place for working class men to hang their votes.

Full social inclusion of women, racial minorities, homosexuals, people of non-Christian heritage, and analyses that thrust environmental considerations into decisive economic dynamics—all this arouses the uneasiness, scorn, hostility, and contempt of many men (especially, but certainly not only, white men) who feel their normative and conventional prerogatives being squeezed, challenged, or threatened. Hence (for instance) the huge support for Bush and other right-wing politicians pro-

1. Thomas Frank's *What's the Matter with Kansas?* is an interesting exploration into "how conservatives won the heart of America."

vided by nonunionized working-class white men: men with flag decals, NRA stickers, and football jackets who hunt, fish, smoke, chew, drive four-wheelers and snowmobiles through the woods, drink, and watch professional sports on TV—Real Men who hate abortion, drugs, gun control, illegal immigration, and homosexuality. At the core of the reactionary male status quo is a combination of violent (or potentially violent) physical strength and a psychological/cultural capacity, with ancient history, to exercise that strength in overpowering ways. Women have traditionally deferred to this strength, and the Left as a whole has not been able to climb out of the deferment pit.

The question, at one level, becomes this: has the Left demographically bumped up against the gender and culture limits of its political viability? Or do the present conventional obstacles and gender-identity barriers represent a log jam about to break open, a breaking that will expand the range, influence, and power of the Left beyond any present expectation? Has the Left reached the boundary of its influence? Or is it about to flood the world with a sea change of a gentler and more inclusive political consciousness? The answer, it seems to me, hinges on the alleged immutability of conventional male imagery and power in the areas of religion and politics, above all. Do the women's movement, reformist environmentalism, the effort to achieve legal parity on the part of racial minorities and same-sex partners represent only a filling-out of remaining political pockets of self-interest within the limitations of the existing system, or do they reveal a bubble about to burst?

We certainly can identify certain factual developments that point beyond the straining status quo. Chief among these developments is the deadly, unsustainable nature of the status quo itself. If, in broad outline, following the dynamics laid out especially by the late Lewis Mumford, we identify the prevailing status quo as the global culmination of civilized institutions of governance and control instituted roughly five thousand years ago, what is unmistakable is the deadly toxicity of its present weaponry, the reach of its economic power, the invasive sophistication of its technology, its capacity to disturb if not totally wreck all previous cultural forms, and its deepening and spreading ecological impacts. Civilization has not only become fully globalized, it also (fully in accord with its imperialistic heritage) exerts its ruling hubris without compassion or remorse.

Next, we can plainly see that various factors associated with the global congealing of this present status quo—the aftermath of European expansion and colonialism, incredible technological and scientific developments, the factual discrediting of religious-based "history"—have served to bifurcate religious sensibility. On the one hand, those who accepted, for instance, the hugely larger time frame of scientific astronomy, geology, biology, anthropology, and archeology have felt compelled to reformulate and reconfigure religious meaning, in light of these scientific discoveries and disciplines, by a careful exploration of the ethical core of religious mythologies. On the other hand, those who refuse to accept the larger time frame, because it contradicts absolute scriptural authority and is therefore undermining of theistic mythology, are obviously feeling their "truth" under cultural threat and are rapidly deploying their answer to that threat: End Times scenarios as "predicted" in various scriptures rich in complex symbols, in some instances the literal End of the World with the theistic God bursting in on an oblivious and unsuspecting humanity to crush unbelief and unbelievers once and for all. That is, from within the most concentrated forms of conventional religious mythology and imagery come increasingly agitated assertions regarding the End of Time and the disasters that announce or accompany the End of Time. If utopian civility, in the military arms of governance, has the factual capacity to eradicate all higher life forms on Earth, its religious counterpart in its most strident articulation is proclaiming that the End of the World is at hand, just ahead, right around the corner. We might even say that fundamentalism is to spirituality what utopian civilization is to human self-governance.

If women's intensive gathering skills created horticulture, and if horticultural abundance opened the way for the domestication of animals and agriculture, and if agricultural abundance produced such a new level of stable wealth that it "invited" an armed aristocracy to expropriate and control this wealth, then we already have deep cultural dynamics that place gender in a powerfully complex relationship to class. If the divinities of the early agrarian village were largely feminine fertility figures—the fecundity of the Goddess—the divinities of male-ruled empires were often bloodthirsty male projections requiring blood and body sacrifice. The gathering Goddess got smaller and smaller as the hunting God got bigger and bigger. Hunting itself became the prerogative of an aristocratic warrior class. The peasantry was consigned women's work and rewarded with aristocratic contempt.

Modern technological civilization has brought these dynamics to global crisis—which crisis includes not only such "externalities" as radioactive waste, deadly for thousands of years, but global climate change from the wanton consumption of fossil fuels with their by-product atmospheric pollutants. The ecological cost of utopian civility is now becoming obvious to anyone with eyes to see, just as aristocratic contempt for the agrarian village has resulted in the burgeoning of technological agribusiness and the virtual extinction of the peasantry. In other words, a global crisis is here; we can trace its origins and progress; we know it's going to intensify. Given that intensification, we are back to our opening question: why is it the Left is so scorned, ignored, and powerless, when it has at least begun to address the magnitude of this enlarging global crisis? Why, in the throes of what is most certainly going to become a global crisis of even greater magnitude and intensity, is the Left so weak?

The answer, in one respect, is exceedingly simple: civilized convention has enormous historical bulk and institutional momentum; civilization has gathered to itself (e.g., the Constantinian Arrangement, in the early fourth century, whereby the Roman Empire ate kingdom-of-God Christianity) such additional destructive power and symbolic trappings of power—including even a postured dispersal of power, as in democratic doctrine—that "civilization" has come to be as crucial a concept of attainment for the Left as it is and has been for the Right. Left and Right now argue about what kind of civilization should prevail on Earth. Nobody is permitted to seriously propose the abolition of civilization. The Right's answer to what kind of civilization we should have is disarmingly simple: the only kind of civilization that can prevail is the kind of civilization that is and always has been—a system of interlocking governing institutions that place disproportionate wealth and power in the hands of a few, the political whole of it protected by armed guards from attack both from without and from within, with a fierce, male theistic God watching vigilantly from above.

The Right essentially says to the Left: if you want egalitarian global sharing and economic attainment limited by ecological sustainability, if you want stewardship and servanthood transformed from spiritual to political dimension, then you want something that is not civilization but its antithesis. In the argument thus far, I essentially agree with the Right. But when the Right says this antithesis to civilization can only be bloody chaos and a universal dumbing down of aspiration—savage, primitive, barbarian, heathen, pagan, and whatever other angelic epithets can be heaped upon the withered pinhead of the Left—the Right is confounding its justifying mythology for history and mistaking its self-serving doctrine of depravity and sin for the original blessing of spirituality. In the name of a loving God (who appears to love hard-nosed aristocratic order above all else), utopian civility insists that only militant selfishness can or ever will rule. Either dreadful chaotic primitivity or dreadful totalitarian dictatorship—the Left is blocked by Thomas Hobbes on one end and whistled to a stop by Friedrich von Hayek on the other. It's either the jungle or serfdom. Take your rueful pick.

From somewhere in the bowels of its historic indigestion, civilization belches from the pre-Constantinian spirituality that refuses to be deconstructed by the acids of utopian convention. The Left is in the belly of this belching beast. That is, the Left's essential vision—egalitarian sharing and ecological sustainability—is, at its core, the ethics of a spirituality that refuses to accept elite greed and power as either desirable, inevitable, or justified by God. The Left's vision is, for all its hesitations and contradictions, a vision of a world seriously committed to servanthood and stewardship—a truly humane world—a world no longer in ideological thrall to the why-it-won't-work elite moralisms of haughty assertion and ethical negation. Deference to utopian authority does not have a particularly different emotional texture than deference to immortality-driven religious authority. The Western ideal, in this regard, was the Holy Roman Empire, an aristocratic holy hierarchy with a theistic God both above the earthly hierarchy and also suffusing the earthly hierarchy with sacred character. Though the Holy Roman Empire is a thing of the past (its remnant operates out of Vatican City), the psychocultural dynamic of generic holy hierarchy simply saturates our institutions of church, school, work, and government. The word that most fully represents Holy Roman Empire in our constellation of acquiescence is—civilization.

Civilization has always existed with sacred character and holy hierarchy. It is an entity that cannot exist, or exist for long, without an implicit theism of deference to sacred (male) authority. (Which may help explain the collapse of the Soviet Union.) It is precisely here where the Left is snagged. The Right clearly stands for an explicit and unclouded public merging of theism and governance, and it claims—rightly—that this public merging is in accord with the historic legacy of all civilization. Meanwhile, the atheistic Left often seems to assume that a true democratic civilization must be composed of secular atheists and that a democratic society presupposes an extended and entrenched practical atheism. The problem with the word "atheist" is that its usage, its conventional range of meaning, spills way beyond its etymological mold. That is, an a-theist is someone who denies the conventionally projected *theos*, someone who asserts that the traditional God to whom deference is made, to whom prayers are addressed, does not exist and never has existed: that this God is a figment, a colossal projection, of human infantile consciousness. This denial easily becomes a denial of spirituality, even an identification of all spiritualities as mere garden varieties of the same infantile projection. Christian, Jew, Muslim, Hindu, Buddhist—it doesn't matter. In this view, they're all whacked out on fantasy projection. They're all beside the point, irrelevant, a hindrance, a damned nuisance, in the way, and a total drag on the program of secular liberation, which is democratic civilization grounded in complete atheism.

The problem with this stance is the old one of the baby and the bathwater. For many of us who have, with great slowness and struggle, worked our way to an a-theism from within conventional religion (mainstream Protestantism, for me), we stand ready to affirm that the God of theism is a psychocultural projection that may reach as far back in human consciousness as pre-bow-and-arrow, pre-spear, pre-fire fear of predator beasts (as Barbara Ehrenreich asserts in her marvelous *Blood Rites*). Most certainly this conventional theism has been utilized by virtually every known civilization to project its God-given authority over the subjected populace. But we a-theists do not agree that there is no spirituality, no Spirit. We agree with atheists that God the Bogeyman is a deadly projection of infantile fear, and we are eager to jointly explore how this Bogeyman has shaped human culture and patterns of governance; but we a-theists have tried very hard to attend to the ethical teachings of those who stand at the base of the religious traditions we were brought up in—guided in this undertaking by such exemplary public figures as Tolstoy, Bonhoeffer, Buber, Gandhi, King, and Day, all of whom were fully and explicitly spiritual. We find there a dimension of being that is below, above, and beyond rationality, a dimension that we freely and willingly call spiritual. We also affirm that this spirituality is infinitely and renewably compassionate, that it trusts and is willing to risk trusting human decency, that human order has an organic character to it, is essentially self-regulating, and that no governing aristocracy (of whatever class origin) is necessary for the continuation of this organic, self-regulating human decency. We are convinced (and our deepened experience confirms our intuition) that to love our neighbor as we love ourselves, even as we do this poorly and inadequately, opens our spirits to the ineffable tenderness of Spirit, and that an identity scarcely dreamed possible is opened by virtue of the cultural realization of stewardship and servanthood.

We a-theists, therefore, in spiritual trust, take the next logical step. If civilization, aside from all its false and flattering attributions of art, literacy and refinement, has been the key historic agency for human oppression and ecological degradation, we stand ready to proclaim the necessary abolition of civilization. Our trust in Spirit and in human decency runs that deep. But so long as the Left insists on the hopeless task of recreating civilization from the Left, the Bogeyman of Power will win. When the Left is truly ready to trust human decency, ready to embrace decentralized ecological community, ready to abandon violence and military power, ready to recognize an equal and equally deserving humanity in every human person irrespective of sex or ethnicity, ready to truly live in Creation rather than merely off it, then the hesitation, doubt, uncertainty, vacillation, and wavering of the Left will evaporate, be over, and masses of ordinary human beings, recognizing the obvious and wholesome spiritual conviction of the Left, will abandon in droves their tentative allegiance to the calculating and totally insincere authority of the mythologically infantile Right.

To recognize utopian civilization for what it is, to renounce and abandon its practices, may seem the ultimate folly, a masochistic choosing of powerless victimhood. I suggest just the opposite. Eutopian spirituality is the doorway into a genuinely humane world, a real culture of stewardship and servanthood, a re-embeddedness in Creation. More than that: given the horrific magnitude of the deepening and enlarging global disasters brought to crisis by predatory civilization and its mighty male religious mythologies, such renunciation and abandonment is the only way out of the theistic apocalypse the infantile Right is busily concocting. Utopia, in its global triumph, is about to crack open. Either its cracking will despoil Earth, or eutopia will be born. My insignificant but total preference is for the latter.

22

The Conundrum Green Politics Is In

The purpose, the drive, and the goal of Green politics is to encourage and create the conditions for ecological living on Earth. To have a grasp of ecological integrity or coherence, it's necessary, to some extent, to rely on science. Scientific studies have essentially convinced us that there is such a phenomenon as Global Warming, that Global Warming is caused by human beings unleashing huge quantities of fossil energy through the devices associated with the industrial revolution, that Global Warming is already resulting in global climate change (with more to come), and that climate change is going to have a huge impact on the economy.

Although Greens rely on the science of ecology to some extent, the desire for Green culture has other sources than science. These other sources are largely cultural. We all have images and ideas of embedded indigenous cultures generally, and of Native American cultures specifically. We have some variable grasp on peasant village culture and the stable culture of small-farm communities. Many of us have a longing for a far deeper-lived connectedness to nature, a life of ecological community, shaped to some extent by these images and ideas. Greens, like all human groupings, have developed a spectrum of adherents. At one end are what I'll call Green yuppies. (Most of these Green yuppies are probably in the Democratic Party, still hammering Ralph Nader because Al Gore lost—or was unconstitutionally deprived of—the 2000 election.) These Green yuppies stand in favor of two things primarily. One is Green technofix for the increasingly obvious dysfunctionality of the economic infrastructure. The other is population control—which increasingly includes restriction on Third World immigration to the First World. In the language of the more politically experienced European Greens, these are the "Realists."

I'm honestly not sure whether there is much of another end to this spectrum. In Germany, the spectrum's other end was called

"Fundamentalist" because it resisted technofix compromise and insisted on the elimination of certain technologies (like nuclear power) and an overall major reduction in energy consumption, as well as the reconstruction of rural society with lots of small-scale, organic gardening and farming. It's not that I think "fundamentalists" no longer exist in this country (I identify very strongly with them), only that we've been terrifically marginalized over the last couple of decades. Conventional affluence has seemingly made us irrelevant. Or, to put it differently, one must struggle directly against the thrust of the conventional economy in order to live with any semblance whatsoever of ecological community. This is hard, and few people do it—including environmental activists.

Twenty years ago—thirty years ago—forty years ago—there was a rather large outburst of what came to be called the "counterculture." The word "hippies," of course, is the most common buzzword used to describe this phenomenon. But the sudden appearance and excessively quick blossoming of the counterculture is, and probably will remain, a little mysterious. Nobody really knows what caused it, even though there are clear causes. Among the causes are these: a current of critical protest against rampant industrial aggression, against racial segregation, against women's enforced domesticity, against the abusive treatment toward Native Americans, against the cultural and economic inequalities of the class system in which the economy functions, against willful ecological desecration, and so on. Dead center in all this was the criminal war against Vietnam. This current of protest against The System runs back easily as far as Thomas Paine; is certainly there in Thoreau, Emerson, and Whitman; is overtly there in Elizabeth Cady Stanton and Sojourner Truth; forms a political party in the late nineteenth century (I refer to the People's Party); generated Eugene Debs, Emma Goldman, Carl Sandburg, Norman Thomas, Malvina Reynolds, Pete Seeger, Rosa Parks, Malcomb X, Martin Luther King, and Ralph Nader. The hippies were preceded by the beatniks, who were preceded by the bohemians. All these were people with something of an artistic bent or contrary ethical temperament who somehow missed, resisted, or declined the compulsive middle-class training in respectability, security, and conventional taste that runs so totally rampant in this decultured society.

The Great Depression of the 1930s underscored the ineptitude of capitalism to fit and respond to human need. World War II provided what Seymour Melman has called "the permanent war economy" on which—I want to underscore this point—the capitalist economy continues to float and is protected, nourished, and guided by. The war economy of the 1940s not only dissolved the Great Depression; the stimulus of subsidized production for mass destructiveness has never been curtailed. Since the early 1940s, we have indeed lived within a "permanent war economy," and we have no way of knowing whether the Great Depression lies waiting in the basement of the military-industrial complex. (I suspect it is. I further suspect the only real solution to the pathological irrationality of utopian industrialism is Green socialism.) World War II also gave us Auschwitz and Hiroshima. It gave us the ruin of overt European colonialism and the steady spread of American hegemony, now essentially unchallenged since the collapse of the Soviet Union. Instead of overt colonialism, we now have the World Bank, the IMF, "free trade," GATT, NAFTA, and the WTO.

Something happened in the last 150 years, more or less, that never happened before in all human history. Well, lots of things have happened that never happened before. Among those many things are plastics, a great variety of chemicals, nuclear power, genetic engineering, cloning, robotics, travel through the air, satellites circling the Earth, instantaneous voice communication via wire and electrical impulse, computers. But the force that lies behind all these particulars is the unprecedented *global* hegemony of civilization. What's never happened before in human history is that civilization rules supreme over the entire planet. If it has not totally extinguished noncivilized cultures, it has rendered these noncivilized cultures powerless, left them bewildered, disoriented, and broken. The civilized economy has everywhere disturbed, and in many places simply obliterated, the noncivilized economy. The civilized industrial economy, fostered and protected by state power, has wrecked the ancient subsistence practices that were the true root of human economic history.

The ecological crisis, if one stands back and looks at the historical unfolding, is the shadow of civilized supremacy. The civilized economy is trying mightily to totally supplant the noncivilized economy. Its symbol is the proliferation of terminator seeds. The cost of this supplanting is the wreckage of noncivilized culture, the imperialist imposition of technological infrastructure worldwide, the total disregard of spiritual restraint, and ecological havoc. This pattern of disruption is true for prior civilizations as well, but on a far smaller and enormously less toxic scale. Therefore the ecological crisis, when traced to its root, is caused by civilization. But we are, as a society, prevented from arriving at this obvious conclusion because the worship of civilization is our real religion. Progress is one of our gods. Economic growth is another. All religions also require an image or a depiction of evil as a way of highlighting or underscoring what's good. In the utopian religion of civilization, progressive economic growth is good, but evil lies in "backwardness," in "primitivity," in "paganism" in its most basic etymological sense.

Enter Green politics. Now what does Green stand for? What it most fundamentally stands for, to repeat my opening sentence, or a portion of it, is to encourage and create the conditions for ecological living on Earth. That is, Green *politics* is really about the creation of Green *culture,* and the promotion of Green culture immediately thrusts us into the direction that civilization defines as evil, a movement toward a reintegration with nature—or, as I prefer, Creation—and this movement toward nature, into Creation, is of necessity a movement into reduced consumption, true conservation, and the recovery of some forms of subsistence and self-provisioning culture. Green contains complex elements of cultural retrogression. Green politics is up against a bigger wall than socialism ever was, for the bulk of socialist agitation and demand was only for a more equitable sharing of a civilized and industrial economic pie that was (so it seems) always getting bigger. Green politics has the unenviable task of asserting that the civilized economic pie is in many ways toxic and carcinogenic. But to make this assertion in the electoral arena is an invitation to a buffoons' ball, a position not unlike that of virtually every prophet the world has ever known—an invitation to be mocked, to be made fun of, to be blamed and scapegoated, and to be ignored. Nobody wants to go "backward."

The extent to which Ralph Nader has been blamed, mocked, scapegoated, made fun of and ignored is a case in point. That he is blamed for Gore's loss in 2000 and Bush's win is a sobering spectacle. The Supreme Court is, I'm sure, amused—by a vote of five to four. The so-called "Realists" sucked up to the Gore machine because *there,* they believed, at least was an opportunity for pressure and influence. And, no doubt, within certain boundaries, that probably is true. I also tend to believe Gore would've been, in some ways, a less vicious president than Bush. But was "less vicious" enough of a reason to vote for Gore? I voted for Nader, and I would vote for him again today, even though I've never truly felt that Nader understands or is able to convey the urgent necessity of creating Green culture. The conundrum Green politics is in is precisely this—that the bulk of the citizenry will resist a reduction in "prosperity" so long as there are no Green cultural models to rely on, and the longer "prosperity" grinds on, the fewer Green models there will be. Yet Greens have no option but to insist on reduced consumption. Therefore, Green politics is boxed in. On the one hand, very few Greens live in anything resembling ecological community. On the other hand, to wait for environmental and economic pressure to *force* the citizenry toward the Green analysis and program is risky. Such pressure can also drive voters to choose the most reactionary advocate for conventional prosperity. Witness the George W. Bush administration.

Now if Al Gore was a chartreuse military-industrial yuppie, Ralph Nader is a political monk. Nader has fought Powers and Principalities all his adult life. He is a knight in the smoke-filled back rooms of dragons' dens. He wears the same rumpled suit he was born in, and the same crooked tie. Ralph Nader isn't a gardener. He's not a midwife or a farmer. He's no stone mason or cutter of firewood. He's never built his own log house. He doesn't live in a commune or on a cooperative farm with large gardens, a wind generator, fishponds, horses and buggies. To my knowledge, he doesn't play a musical instrument, sing, or recite poetry. Please understand: I am not criticizing Ralph Nader. He is a fantastic man, a wonderful human being, a blessing to humanity. But whether or not he knows anything about the urgency of Green culture, all he projects is Green politics. We have no comparable model for Green culture—no Gandhi to unite culture and politics. So, at this stage, socialism can gain more support in the American electorate than Green politics. Witness the growing public interest in universal ("single payer") healthcare. Therefore, let me conclude this essay by saying a few good words about socialism.

There are two pillars, two foundations, of ethics the world over. One of these is stewardship in Creation. The other is servanthood among our own kind. Socialism, for all the bad rap it's gotten, a lot of which is fully deserved, is essentially built on the second foundation of servanthood. Socialism is about sharing, earnestly so. Green, on the other hand, is about stewardship in Creation. It's about living as lightly on the Earth as possible, as poison-free as possible, as respectful and sensitive to the rights of all creatures, all landscape, all of the natural world as possible. Stewardship and servanthood are twins who need each other. Separate, they are destructive. An industrial doctrine of sharing without ecological restraint and without respect for ecological community results in the utopian disaster of Soviet communism. Stewardship without sharing results in the exquisite gardens of aristocracy, while peasants starve beyond the lovely stone walls. Green cannot be truly ecological without sharing. Socialism cannot be truly sharing without stewardship.

The task at hand is for these ethical twins—sharing and stewardship, socialism and Green politics—to discover and cherish their common humanity. Only with their mutual trust and reciprocal respect will there be a political program that begins to address the depth of the crisis we face. We need socialized mass transit and we need the ecological, self-provisioning commune—Red needs Green and Green needs Red—and the sooner we get at it the better.

The alarm is going off on the ecological clock. The question is urgent and becoming more urgent. What must we do to wake up?

23

The Vision Thing

To not only be content with, as an ecological and ethical necessity, but to actively encourage and embrace an ecologically coherent way of life, is at once theoretically attractive and practically repugnant, a "going backward" rather than a "going forward," a "cutting edge" idea of the "simple life" we prefer to daydream about rather than engage. We can satisfy our guilty consciences in regard to global poisoning by shopping for organics at a farmers' market, global sweatshops by buying through Co-op America, global warming by purchasing the most fuel-efficient car. And while being (or becoming) a better consumer is not without ethical merit, an ecological society is not a cultural configuration arrived at through market contrivance. The market, in fact, can readily accommodate all sorts of niche consumers, with total indifference as to ecological or humanitarian virtue. The model we need is a model discredited by our deeply-ingrained civilized bias. That model is the peasant village.

Let me hasten to assert that I do not propose the peasant village as a model for ecological living with any sort of uncritical finality. There are several factors that make the peasant village only a rough model, and one to be sifted through with patient care. At the historical base of this peasant roughness lies an ancient history of oppression and expropriation by aristocratic civility. This means that the subsistence and self-provisioning practices of peasant experience under aristocratic control have been rooted in the thin soil of an enforced scarcity. The cultural options of the peasant village have always been limited, and those limitations have been enforced by elite violence and the threat of violence. What the peasant village could or would have evolved into, had it not been colonized by a nascent bandit aristocracy at the "dawn" of civilization, is fascinating but speculative. This question epitomizes the thrust of the five epigraph quotes that open this book, most fully characterized by James Carroll's remark that "The study of history always implies a study of its alternative."[1] The chief asset of such speculation is that it teaches us not to accept the impoverishment of the peasant village as natural history, as if it were impoverished by divine decree or a law of nature. Civilization's dawn is the twilight of the free village.

We have been so—pardon me—brainwashed by the prerogatives of aristocratic civility, so dismissive of the culture of folk life, so glad that our lives have been freed from the hard, earthy drag of unrewarding agricultural economics, that our hardest moral tasks are to (1) think critically about the predatory core of civilization, and (2) reconsider the cooperative and subsistence model the peasantry holds for our future. That we find it so incredibly hard to think along these lines—so full of gratefulness for liberating civilization, so indifferent (when not overtly hostile) toward the imprisoning peasant experience—is a huge part of the reason why we remain so politically incapacitated and morally bewildered, so ethically numb, in the face of what has openly become a brazen, global, civilized assault on both ecological integrity and cultural coherence. If civilization is our savior from "barbarism," "terrorism," "primitivity," "paganism," and all the other dreadful monsters that lurk beyond the skirt of mercury vapor light, how can we ever bring ourselves to think critically of that which (we've been taught) saves us from such danger and destruction, and is our salvation from both the spiritual and material consequences of sin? The thing we most need to consider and explore—cooperative ecological living—is, if not the farthest from our minds, at least the farthest from our actual intentions and plans.

The Left talks earnestly about sharing the largesse of the economy. This is, in principle, an absolute necessity. We *should* have universal health care, a minimum (and a maximum) income threshold, as low-energy a form of mass transit as is possible, including (why not push for a positive aspect of NAFTA?) a Canadian, U.S., and Mexican train service called Camex, far more leisure, vacations and holidays, time to meditate, contemplate, play, and pray. But the bulk of the Left derives its world-view, or, more exactly, its *economic* world-view, from utopian civility. With the exception of that strain of anarchism represented, for instance, by Peter Kropotkin, Martin Buber, Paul Goodman, and E. F. Schumacher, the Left has had an ideology of maximized industrial progress that is, in many ways, identical to predatory capitalism. From fields to factories to computers, the progressive instrumentalization of the natural world and the methodical elimination of those cultures that resisted that instrumentalization have been the track records of both Left and Right.

This instrumentalist world-view simply saturates our consciousness. The computer, the video game, the television, the supermarket, the strip mall, the medical infrastructure and procedures, the sporting events, the transportation systems, the educational curricula and structure of compulsion, the military budget and its practical embeddedness in huge areas of human life and endeavor, the energy-producing systems, the prison industry, the space program—all engulf us in their comprehensive hegemony of utopian achievement. All else, all that came before, has merit only insofar as it tickles our fancy. We don't *need* the past. We don't *need* noncivilized culture. The peasant village is a log we've walked across to get to the Promised Land, and we've entered ("the information superhighway into the twenty-first century") that Promised Land, having slipped past the Y2K trolls. Those who deny it are, at best, romantic misfits in need of an attitude adjustment. Probably they are terrorists whose psychological profile is being studied at the appropriate level by joint government-university research. The hard core will find a camp for sustained concentration at Supermax. The ramifications of our psychological and spiritual distortions, while not themselves psychological (Global Warming is devoid of psyche, unless we mythologize the dynamics by saying Gaia has a fever), are not to be adequately addressed, however, without plunging into the human distortions that created the undesirable ramifications in the first place. Technical fix, at some level, is, if possible, a necessity; but technofixers devoid of soul are only playing video games with disaster.

The only truly positive thing to say about civilization, and even then the positive thing is an unintended consequence, is that its global hegemony in disaster is forcing humanity to the wall, forcing us to counter the rush toward universal toxicity by facing into our accrued distortions. Some of these distortions are explicitly civilized: massive slavery and institutionalized warfare, a sustained cultural (and power) cleavage between aristocrat and peon, between First World and Third. But some distortions are precivilized or noncivilized in origin, of which male presumptions of power, including ritual violence, are critical. These presumptions have been and still are embedded in our functional life-definitions of gender: little girls with dolls, little boys with toy weapons. This precivilized or noncivilized pattern of male prerogative has fueled the conduct of civilization; it is the core around which civilized institutions, including the economics of civility, have evolved. The history of civilization is the history of compulsion and male violence.

It's this history—this embeddedness of compulsion—that makes women's liberation such a terribly complicated paradox. Insofar as women want equality within the civilized structure, without a fundamental critical reappraisal of that very civilized structure in which they seek to advance, the tendency is toward taking on the prerogatives of male power, of concurring in a system, however rationalized, of economic compulsion and expropriation. Women who do this, particularly in politics, business and the military, rise the fastest and become, *ipso facto*, ambiguous role models. The true gender issues to be resolved seem to me to be these: men must abandon the presumptions of traditional male prerogative, of which resorting to violence or intimidation is high on the list of particulars; women must openly confront the predatory structures of compulsion that simply saturate civilized economics (and subsidiary institutions like schooling) and consent to a kind of "going backward" that is the critical, ecological ramification of Green culture. But women, if at all possible, will *not* consent to a "simpler" lifestyle in an ecological economy if men do not renounce, abandon, and repent of their accrued and habitual power presumptions.

With some 32,000 nuclear weapons in the world ("at last count," according to Jonathan Schell[2]), this is the impasse in our extremity. To say the situation is not extreme is both false and absurd. We live, literally, in the shadow of extinction, every minute of every day, and civilization, with its organizational core of male power prerogatives, has brought us to this seemingly exitless shadowland. Eutopia versus utopia may be an odd and somewhat eccentric way of expressing the dynamic we are in, but it is fully functional and apt. Green politics and Green culture, backed by assertive feminist values and a choir of Rainbow voices, is the only viable pathway out of the toxicological power box we've become trapped inside of.

The peasant village, free from extortion and expropriation, is the model from which we can pick and choose. This is the late Stone Age garden path from which humanity was diverted by the sustained, organized banditry of civilized predation. Contrary to our manufactured fears, ecological living in nonviolent community offers and contains a magnitude of human liberation only waiting to be discovered—an earthiness, a creatureliness, that will vastly enlarge and truly enhance our humanity.

It the meantime, we are like the dwarfs in *The Last Battle* by C. S. Lewis: already beyond Narnia, we huddle with eyes squeezed shut, fiercely determined not to be "tricked." For the time being we prefer, in our civilized comfort and convention, the illusion that "deterrence" protects us from the disasters that are rising up around us. Our comfort makes breakdown all so "hypothetical" or (as in the case of terrorist attack) so unlikely because of "homeland security." Is it only pain and anguish and grief that will cause us to open our eyes? Is suffering the fulcrum of political leverage? If so, we don't have long to wait.

NOTES

1. Carroll, *Constantine's*, 17.
2. Schell, "New," 11–15.

24

Preface to Fearful Energies

It's possible to conclude, from the language of the preceding essays, that I am merely indulging in semantic absurdity and concept mongering; and, no doubt, there will be readers who think just that. For some, there may be nervous puzzlement as to whether civilization-as-utopia or peasantry-as-garden-id is simple word play, poetic lightheartedness, or whether these and other formulations reach into a psychocultural depth that sheds light on the underlying *value* dynamics of our present deadly malaise. For others, these ideas, these articulations, may simply ring true and bring fresh possibilities for conduct and action in the dimension of emerging Green culture.

Norman O. Brown says the Freudian challenge is to bring the unconscious into consciousness. I would only modify that challenge by adding that the task is to bring *more* of the unconscious into consciousness—for the simple reason that, as biological beings, we can never achieve (nor should we necessarily *want* to achieve) total consciousness. (Total Consciousness is a basic immortal heresy of utopia, anyway.) There is an implication here, and perhaps a meaningful one. Bringing the unconscious into consciousness is, as scholastic philosophy used to put it, necessary but not sufficient. It is only "on the one hand." The other hand is this: as the unconscious enters consciousness, consciousness begins to feel saturated with the here-and-now (the immediacy of reality and life) and easily gets swamped. The present moment is intensified by "id" energy from the unconscious. This opens, as Aldous Huxley said, the doors of perception and the gates (perhaps the floodgates) of feeling. It is this here-and-now intensification that conventional consciousness feels uneasy about and tries (often successfully) to prevent or suppress; the political climate in regard to the "drug culture" is an ambiguous example of current familiarity.

Perhaps the counterculture as a whole is a useful illustration of the ambiguities and contradictions in "bringing the unconscious into consciousness," for hippies were by upbringing largely middle-class utopians whose consciousness dipped into eutopian visions, for the most part unsustainable, because the necessary nonurban cultural skills were lacking, along with the spiritual discipline, by which to sustain eutopian consciousness and thereby build a truly rich and stable Green culture. Hippies were all too willing to live off the fat capitalist cow, not weaned from utopian effortlessness, and lacking the cultural memory or principled spiritual conviction that would have informed them that decadent imperialism extrudes a shoddy self-indulgence in its affluent bloatedness.

Drugs are not recreational, in the usual sense of the word. They were used recreationally in the hippie counterculture in inverse proportion to practices of spiritual discipline. Which is to say, "id" energy *does* generate foolishness and disruption when it floods undisciplined or unprepared consciousness, especially when stimulated by potent mind-altering chemicals. The implication is clear: a eutopian world presupposes a simultaneous *spiritual* deepening as well as deeper bodily awareness as biological beings. To bring *more* of the unconscious into consciousness demands greater *consciousness*, more intentionality, and a keener engagement with an ethically-responsible spirituality. Therefore, we stand in need (as the counterculture stood in need) of a sustained communal spirituality, with meditative and psychotherapeutic capacities, that continues to integrate fresh "id" energy into communal cohesion.

There are reversals, ironies, dialectical tensions, and outright contradictions in these intellectual mappings of psychohistorical cultural transformations. Because the map is not the territory and the menu not the meal, we realize from the outset that language is not the reality pointed to. But neither is language outside of reality; if it were, where would it be? So the first dialectic, if it is the first, throws "subjective" description and "objective" reality into perceptual tension. This is essentially the turf of art, which implies that Green culture will be rich in music, sculpture, painting, crafts, poetry, and arts of all sorts. We will get our voices back, as we already are, in music, poetry, speech making, and deepened conversation. Anyway, what I set out to say in this little preface (before getting happily lost in garden lanes) is this: for those who have the tolerance and maybe even the aptitude for ironies, reversals, correlations, reverse correlations, dialectical tensions, outright contradictions, and the general muddle of trying to sort things out—here's more.

But one final note. I am obliged to use "utopia" in quotation or reference (e.g., Martin Buber's *Paths in Utopia*) even when I am convinced that the author means "eutopia," as I believe Buber does. This causes confusion. I will do my best to minimize the confusion, but I cannot just undo it—which proves, I think, that utopia and eutopia are due a dialectical healing, too.

25

Their Fearful Energies

CIVILIZATION, BROUGHT TO A new pitch by scientific discoveries and technological innovations, in the context of a massive (and growing) world population, deepening poverty and ecological instability, is in an explosive state, a sharp disequilibrium over against both nature and non-civilized culture. Efforts to defuse this explosiveness are desirable (who among us wishes for endless terrorism or atomic war?), but we need to examine as carefully as possible the implication of centralization, on the one hand, and decentralization, on the other.

Is there a transcending purpose in civilization, despite its obvious and destructive globalizing excesses, its traumatic institutions? One must assume that there is such a purpose; perhaps material progress, even the destructive dynamism of the industrial period, contains within itself the hidden agenda of suppressed transitions. The most earnest messengers are not necessarily aware of the message that they carry. There are profoundly "civilized" reasons to be in sympathy with the ideals of democratic socialism and cultural internationalism; and these ideals clearly imply certain levels of centralization: the United Nations, for an obvious example. Internationalism is the logical result of both a tough ethical gentleness and that unifying process of culture of which the city has been the leading agent. However, an internationalism that destroys communal integrity and regional authenticity also destroys the foundations upon which internationalism must rely. In these essays, I am stressing the need for balance and integration: between the city and the countryside, between industry and agriculture, between spontaneous education and organized schooling, between organic health and clinical health care, between the fabricated and the natural, between the individual and the communal, between the designed and the eolithic, between civilization and whatever we might call its opposite (folk culture, perhaps), and—not least—between women and men.

It is my contention, built above all on the historical and intellectual constructions of Lewis Mumford, that the single greatest achievement of modern urban-industrial-scientific centralization has been the growing liberation of women and the public empowerment of the feminine. This liberation, despite all the cultural ambiguities, is real. Yet the cost of this advance, in terms of contemporary human suffering and environmental damage, is colossal. But it is necessary to stress that this destructiveness is not caused primarily by the women whose freedoms are rapidly growing, but by the men whose policies and institutions are predatory and rapacious. It is simply a demographic fact that the political, military, and economic institutions of modern industrial society are overwhelmingly *male* in structure and execution. To a large extent, women, as they increasingly enter the world of previously all-male professions, have no choice but to engage these institutions exactly as they find them and, in the short term, to accept them as they are. This leaves women open to the charge, as it does also in regard to ethnic minorities, of participating in an uncritical integration—reflected in the tougher-than-men atmosphere of Margaret Thatcher, say, or Madeline Albright, or Condoleezza Rice.

The political liberation of women has, however, grown simultaneously with the excesses of industrialism and urbanism. (In the United States, women won the vote almost exactly as urban population began to exceed rural population—1920.) Yet it is as if the steady advance of women toward the fraternal power centers has aroused an intensification of the competitive struggle of man against man: an unconscious acknowledgement that the very principles of male action, sacrosanct within a closed fraternal mentality, are now themselves being penetrated by feminist assertion, analysis, and demand. As the intellectual justification for traditional gender roles has been demolished by science and empirical data, and as the need for gender roles has been largely subverted by technology, the underlying male bias is revealed in its barest form. Their outer fortifications of prerogative having been destroyed under a barrage of ruthless objective scrutiny, men now huddle in the only room left—the powderhouse—threatening to blow up everybody (including themselves) unless the advancers back off.

In this context, a call for decentralization could be seen as a call for retreat, a historical regression. It could give men a chance to reformulate their strategies, devise new options, to retain their preponderance of power. Decentralization is critically important to ecological survival: life without an earthy indigenousness succumbs to technological sterility. But until women achieve full participation in political power on the basis of complete equality—an equality that reaches into arenas of control, notably "defense," that remain the exclusive preserves of men—the first issue will remain the sharing of power and policy-making *at the center,* not the rebuilding of life from the circumference. The advance of women is irrevocable, and the bluff of those in the powderhouse is being called. This may seem psychoanalytically mad (such profound transformations of consciousness and culture are bound to appear mad in the throes of change), but the moral weight lies with the advancers, the women, and it would be mass cowardliness to call off the advance in the face of genocidal threat. Are women, having finally cornered men, going to retreat because men have their fingers on the triggers of long tubes whose heads, on impact with Earth, would explode in unimaginable destructiveness? The elemental symbolism is so obvious that only the most ideologically blind could fail to see it or appreciate the tragic gallows humor.

There are ways to defuse the crisis, but the only honorable way is to face the issues with humility and courage. Reconciliation between women and men means, in part, a redistribution of power: the inclusion of women on an equal rank with men within the highest circles of government *would* result in a significant shift of policies, an amelioration of military confrontation. Men, left to their own counsel, are not going to disarm, nor will they significantly reduce the rate of economic growth (if they can help it) even if it means continuing to embrace the stupendous hazards of nuclear reactors with their apparently unsolvable waste disposal problems. Indeed, the prospect of a world without constant economic growth is for many men a vision that engenders psychic terror. The law of civilization is expansion, and the blockage of expansion spells decay. To be sure, there are men who counsel differently. Not all men and women are queued by sex in opposing lines of confrontation. But men who advocate either the reform of present institutions or their radical transformation, yet who do not grasp the depth of the gender issue, are, with their partial solutions, also a part of the problem. Ivan Illich, in *Deschooling Society*, represents a case in point.

II

In Illich's book, in a chapter called "Rebirth of Epimethean Man," there is a painful example of an honest approach to, but final avoidance of, the problem. Illich says

> The original Pandora, the All-Giver, was an Earth goddess in prehistoric matriarchal Greece. She let all ills escape from her amphora (*pythos*). But she closed the lid before Hope could escape. The history of modern man begins with the degradation of Pandora's myth . . . . It is the history of the Promethean endeavor to forge institutions in order to corral each of the rampant ills. It is the history of fading hope and rising expectations.[1]

This distinction between hope and expectation is one of Illich's excellent insights. He says that hope means "trusting faith in the goodness of nature, while expectation . . . means reliance on results which are planned and controlled by man." Illich goes on to say that "The Promethean ethos has now eclipsed hope. Survival of the human race depends on its rediscovery as a social force." Clearly, in this analysis, hope is a feminine attribute, while expectation derives from the "satisfaction from a predictable process" of Promethean construction, male in origin.[2] (With some imagination one can trace these attributes further back in time: the prehistoric woman may *hope* that the man will return with meat; but the man, willy-nilly, *expects* to be fed when he comes back to the hearth. Thus the Promethean ethos attempts to systematize very archaic demands, demands that may have their origin in the infant at his mother's breast.) But let us go on with Illich's analysis:

> The Greeks told the story of two brothers, Prometheus and Epimetheus. The former warned the latter to leave Pandora alone. Instead, he married her. In classical Greece the name "Epimetheus," which means "hindsight," was interpreted to mean "dull" or "dumb." By the time Hesiod retold the story in its classical form, the Greeks had become moral and misogynous patriarchs who panicked at the thought of the first woman. They built a rational and authoritarian society. Men engineered institutions through which they planned to cope with the rampant ills. They became conscious of their power to fashion the world and make it produce services they also learned to expect. They wanted their own needs and the future demands of their children to be shaped by their artifacts. They became lawgivers, architects, and authors, the makers of con-

> stitutions, cities, and works of art to serve as examples for their offspring. Primitive man had relied on mythical participation in sacred rites to initiate individuals into the lore of society, but the classical Greeks recognized as true men only those citizens who let themselves be fitted by *paideia* (education) into the institutions their elders had planned.[3]

It should be said, at this point, that the essential intention of the primitive men and classical Greeks of whom Illich speaks is identical. In each case, the purpose of the men is to initiate boys into manhood, to give them a male-controlled "second birth." The major difference between the "sacred rites" of the primitives and the *paideia* of the Greeks was in the nature of their respective institutions: the former acted by word of mouth, by lore, and by mythological participation; the latter acted by the written word, by history, and by concrete planning. But let's go on with Illich's narrative:

> By stealing fire from the gods Prometheus turned facts into problems, called necessity into question, and defied fate. Classical man framed a civilized context for human perspective. He was aware that he could defy fate-nature-environment, but only at his own risk. Contemporary man goes further; he attempts to create the world in his own image, to build a totally man-made environment, and then discovers that he can do so only on the condition of constantly remaking himself to fit it. We now must face the fact that man himself is at stake.[4]

Yes, *man himself is at stake*! This is the real issue in its nongeneric brazenness. Illich, recognizing that the Promethean ethos has produced a global cul-de-sac brought about by the constructs of "misogynous patriarchs," calls honorably for its dismantling. Worried about the need of modern man to remake himself, Illich offers us an alternative:

> We need a name for those who love the earth on which each can meet the other .... We need a name for those who collaborate with their Promethean brother in the lighting of the fire and the shaping of iron, but who do so to enhance their ability to tend and care and wait upon the other .... I suggest that these hopeful brothers and sisters be called Epimethean men.[5]

Now I do not wish to denigrate Illich's explicit goals: to call forth those who love the Earth, who collaborate in the lighting of fire and the shaping of iron, and who do so to enhance their capacity to tend and care and wait. These are lovely and tender attributes without which life could not go on. But is it intellectually honest to discard the masculine traits of "Promethean man" only to bring in "Epimethean man" dressed in what have long been recognized as feminine characteristics? One cannot help but suspect the psychology of misogynistic patriarchs has altogether sly and subtle ways of perpetuating itself, even at the cost of an intellectual transvestitism. Would it not (I ask this in all sincerity) be infinitely more honest, since Pandora, we may assume, still has hope in her box, for us all to be called, brothers and sisters both, Pandoran *women*? Illich, needless to say, does not suggest this possibility. It is, obviously, the sort of idea that, if taken seriously, might throw women and men alike into a giddy panic. At the conclusion of Sigmund Freud's *Civilization and Its Discontents*, this same mentality is at work:

> Men have brought their powers of subduing the forces of nature to such a pitch that by using them they could now easily exterminate one another to the last man. They know this—hence arises a great part of their current unrest, their dejection, their mood of apprehension. And now it may be expected that the other of the two "heavenly forces," eternal Eros, will put forth his strength so as to maintain himself alongside of his equally immortal adversary.[6]

The "immortal adversary" here is Thanatos, the death instinct, a late construct of Freud's thought. Although there is no exact parallel between the respective pairings of Prometheus and Thanatos, Epimetheus and Eros, we find in both Illich's and Freud's proposals a softer, more feminine man offered as a new archetype for both male identity and "man." This softer man is to replace the harder man: Eros is to replace Thanatos as Epimetheus is to replace Prometheus. But it's necessary to ask why *all* the "heavenly forces" that both Illich and Freud invoke are masculine.

It is true that a softer, more feminine man is needed if the world is to avoid plunging into catastrophe. Correspondingly, a tougher, more masculine woman is required as women move as decision-makers into institutions of public policy. The shift in masculine identity that both Illich and Freud call for is necessary but not sufficient. To put it in a word: generic "man" has outlived its usefulness. Yet it may be virtually impossible for men to be called generic "woman" without feeling consumed by the feminine. This very fear may well have been the drive in men that created the masculine city (civilization and utopia) out of the feminine agricultural village: the thrust of the smothered sons to throw off the domination of the all-enveloping Mother.

In her book *Knowing Woman*, the late Jungian therapist Irene Claremont de Castillejo says that the "struggle for consciousness is the perennial struggle of the son to break free from the Great Mother."[7] Gerda Lerner, in *The Creation of Patriarchy*, says that much of the "striving for autonomy" and the "recognition of selfhood" arise from the "infant's struggle against the overwhelming presence of the mother." Although she modifies her assertion by saying that civilized society has "interposed itself between mother and child and has altered motherhood," Lerner goes on to say that "under primitive conditions, before the institutions of civilized society were created, the actual power of the mother over the infant must have been awesome":

> Only the mother's arms and care sheltered the infant from cold; only her breast milk could provide the nourishment needed for survival. Her indifference or neglect meant certain death. The life-giving mother truly had power over life and death. No wonder that men and women, observing this dramatic and mysterious power of the female, turned to the veneration of Mother-Goddesses.[8]

Certainly this underlying feminine-masculine tension forms one of the real background dramas against which human history, civilization, and the evolution of human consciousness have developed. All men are born of woman; all must wean themselves from the Mother, turn from her and direct their energies elsewhere. Or, as Gerda Lerner puts it, "In order to find their identity, boys develop themselves as other-than-the-mother; they identify with the father and turn away from emotional expression toward action in the world":

> In civilized society it is girls who have the greatest difficulty in ego formation. I would speculate that in primitive society that burden must have been on boys, whose fear and awe of the mother has to be transformed by collective action into identification with the male group .... The ego formation of the individual male, which must have taken place within a context of fear, awe, and possibly dread of the female, must have led men to create social institutions to bolster their egos, strengthen their self-confidence, and validate their sense of worth.[9]

Lerner points to a basic biological reality that can never be "solved," only grappled with—hopefully with an ever fuller consciousness and deeper understanding. This is the ancient background to Illich's *paideia* and "mythical participation in sacred rites."

The issue is straightforward and profound: women partake of a direct reproductive continuity from which fundamental life satisfactions, unknown to men and only symbolically "reproduced" in "sacred rites," are obtained. Norman Mailer, who has himself been something of a lightning rod for feminist writers, puts the issue simply in *The Prisoner of Sex*:

> So do men look to destroy every quality in a woman which will give her the powers of a male, for she is in their eyes already armed with the power that she brought them forth, and that is a power beyond measure—the earliest etchings of memory go back to that woman between whose legs they were conceived, nurtured, and near strangled in the hours of birth. And if women were also born of woman, that could only compound the awe, for out of that process by which they had come in, so would something of the same come out of them; they were installed in the boxes-within-boxes of the universe, and man was only a box, all detached.

A man, says Mailer, is "alienated from the nature which brought him forth, he is not like woman in possession of an inner space which gives her link to the future, so he must drive to possess it, he must if necessary come close to blowing his head off that he may possess it."[10] And so the drive for equality between women and men occasions an unparalleled crisis.

If men, several thousand years ago, were compelled to throw off the domination of women—even if this domination was merely a perceived one, psychological—so women are now working to throw off the religious, legal, and military structures of male domination, the accrued institutional mazework of utopian civilization. Yet men, aside from all rational conviction, have an organic, deeply-seated fear of being smothered by women. De Castillejo, for instance, says

> Woman's invasion of man's sphere has, I believe, aroused in man's unconscious the determination to maintain his former superiority at all costs, even among those who consciously believe in and are most vociferously in favor of equality between the sexes. Consciously men welcome woman's emancipation but in the unconscious they despise her and are determined to keep her in her place.[11]

De Castillejo's remark may be something of an overstatement; but not much of one. Men *are* afraid of being consumed by women, even when external power is tipped outlandishly in their favor; and men resist political advances by women, not on any particular merit of rationality, but because of powerful, hidden fears. Yet Jungian psychology asserts that both men and women have aspects of masculinity and femininity within them. Irene Claremont de Castillejo says that the Age of Enlightenment set masculine rationality on a throne and devalued the irrational and emotional feminine. "On the whole," she says, "the menace to civilization today seems to come from man's over-developed thinking, and the consequent unadaptedness of his . . . feeling, which thereby lets in the evil."[12] (I would only modify that idea by pointing out that the menace is not so much *to* civilization as *from* it.)

David Bakan, in *The Duality of Human Existence*, writes about "two fundamental modalities in the existence of living forms" that he calls "communion" and "agency." Agency he describes as self-assertion, isolation, and alienation; communion, on the other hand, is characterized by openness, contact, and a "sense of being at one with other organisms."[13] Eventually he identifies agency as "more characteristically masculine" and communion as "more characteristically feminine."[14]

In Jungian terminology, this feminine "archetype" in men is called *anima*; the masculine counterpart in women is called *animus*. Therefore, in Jungian terms, what we are witnessing in the modern world is the dramatic expression of woman's animus, the sudden and forceful exercise of her latent masculinity, and the consequent pressure on men to develop their anima. "So the poor man suffers doubly," writes de Castillejo. "No longer boosted by women on the one hand, and actually competed with by them on the other, he feels depotentiated and unable to rise to the heights expected of him. All this at the same time as he is rightfully developing his own sensitive feminine side."[15]

The complexity of the modern industrial-scientific world, the complexity that Lewis Mumford has referred to as world-as-city, is the nexus out of which these turbulent psychological and spiritual transformations are rapidly gaining momentum through the medium of political, economic, and technological developments. The volcanic city, whose cultural roots and molten psychic magma lie in the archaic feminine village, has now poured itself lavalike over the entire world; and the inner woman, no longer constrained by the outer man, bursts through at this point of utopian consolidation. De Castillejo again:

> The deeply buried feminine in us whose concern is the unbroken connection of all growing things is in passionate revolt against the stultifying, life-destroying, anonymous machine of the civilization we have built. She is consumed by an inner rage which is buried in a layer of the unconscious often too deep for us to recognize. She becomes destructive of anything and everything, sometimes violently but often by subtle passive obstruction.[16]

The "buried feminine" in this analysis refers not only to women but also to the anima in men. The eruption of the feminine is the inevitable result of the compression and compaction of feminine values within a smugly victorious utopian machine. This radical imbalance between the feminine and the masculine (the suppressed masculine in women and the suppressed feminine in men) brings de Castillejo to the point of asking whether this imbalance isn't "one of the deep-rooted causes of the most devastating wars the world has ever known." By asking such questions, one begins to pry into the buried threads of history.

III

Let's try to weave a couple strands of thought from Illich and de Castillejo. The former, in *Deschooling Society*, uses images and descriptions that suggest a body of thought he does not pursue overtly but which nonetheless lies behind his analysis:

> The attendance rule makes it possible for the school room to serve as a magic womb, from which the child is delivered periodically at the school day's and school year's completion until he is finally expelled into adult life. Neither universal extended childhood nor the smothering atmosphere of the classroom could exist without schools.[17]

Irene Claremont de Castillejo says that "boys who are kept throughout their adolescent years seated at school desks in a state of boredom, trying to evade learning matters in which they have no interest, are thwarted of any challenge whatsoever. By the time they leave school the desire to achieve has been stifled in most of them."[18] The "desire to achieve" is an ambiguous description. That there are large numbers of men (and women) who ploddingly consent to the destruction of their creativity and emotional wholeness by allowing their lives to be consumed by industrial and bureaucratic regimentation suggests that de Castillejo is right. The problem with the term, however, is that school regimentation fosters fantasies of wild achievement in children whose *self-directedness* is otherwise thwarted. These fantasies are encouraged by advertising and channeled by the media, bending every eutopian self-directed impulse toward utopian consumption and conformity. But that the school system is the training ground for regimented identity is obvious.

The French cultural historian Philippe Aries, in *Centuries of Childhood*, spells out the increasing contraction of freedom for European schoolboys in the period beginning in the sixteenth century. "We started with a situation," writes Aries, "in which the boarding-school was unknown, and schoolboys lived in lodgings, free of all authority, whether paternal or academic: hardly anything in their way of life distinguished them from unmarried adults." But by the early nineteenth century, a wholly different climate had emerged:

> The development of the boarding-school system after the end of the eighteenth century bears witness to a different concept of childhood and its place in society. Henceforth there would be an attempt to separate childhood from the other ages of society: it would be considered important—at least in the middle class—to shut childhood off in a world apart, the world of the boarding-school. The school was substituted for society in which all the ages were mingled; it was called upon to mould children on the pattern of an ideal human type.[19]

Here again we see that the contraction of freedom for boys, and the corresponding rise of institutional schooling, correlate exactly to the emergence of industrial society, the deepening of civilized infrastructure, the extinction of rural culture, and the systematic imposition of a standardized utopia.

Norman Mailer says that technology, "by extending man's power over nature, reduced him before women . . . ."[20] Mailer's remark occurs in the context of puzzled thoughts on hospital delivery rooms, but two implications can be drawn from his assertion. First, technology in the hands of women reduces their direct dependence on men; second, the social standardization that has come with modern technology has "domesticated' men as never before. When the axe and crosscut saw become obsolete, when an effortless turn of the thermostat "produces" heat, we have an illustration, perhaps even a symbol, of how technology and social organization shape the relations between women and men. The shrinking of traditional forms of gendered work and the simultaneous abandonment of household self-provisioning have stripped away reciprocal economic dependence, made all possible commodities available through outright purchase, and either elevated or reduced the relations between women and men to the dimension of personal attraction, with erotic energy strongly in the foreground. The commercial media are at hand to coach and cheerlead. You've come a long way, baby, and the Marines are looking for a few real men.

Yet de Castillejo says that to be adventurous, indeed to be "heroic," lies near the heart of traditional manhood: "It unquestionably is the male of the human race who is the active partner and has explored the greatness of the earth and the heavens and the smallness of invisible particles. It is also he who became conscious first and has, as a result, been the architect of modern culture."[21] (On the matter of who was "conscious first," de Castillejo seems, at best, to be confusing historical periods. That men have been the architects of modern civilization may well have more to do with power than consciousness. But it *may* be true, in evolutionary terms, that the spirit of unrest that seems to pervade the human male does correlate, in some subtle ways, with first consciousness.) So what is now most heroic—space exploration—dovetails neatly with a regimented social machine: all of the underlying infrastructure from science curricula in elementary schools to foreign policy "strategic interests" that congeals in rocketry, satellites and space stations, all of it embedded in elaborate and intensive "security."

A pivotal point in de Castillejo's analysis is the First World War—a war, she says, that "appeared to have been the result of a masculine culture."[22] The "change in woman's attitude toward the hero in the last fifty years is very marked. Before the First World War it was men's business to be heroes," but in the period "between the two Great Wars . . . women won a great part of their battle for equality."[23] So we see that civilization in its utopian consolidation paradoxically provides the basis for women's liberation. At the same time, extended schooling has dampened the more self-directed "heroic" in boys and channeled it into organized sports and technological fantasies. (Referring to the "heroism of men" in relation to the exploration of space, de Castillejo gives "All honour to their imagination, their skill in wresting the secrets from nature, and above all to their incredible courage. But one cannot help wondering if man's pursuit of the physical moon is not the outer counterpart of his paramount need, of which he is not fully aware, to explore the cold unpredictable half-light of his own feminine nature."[24])

The school, explicit in Illich's terminology of "magic womb," and implicit in de Castillejo's remark about "boys who are kept . . . in a state of boredom," is one manifestation of the Great Mother who, by "subtle passive obstruction," is attempting to contain and soften the unbridled masculine in men. In another passage, de Castillejo says that "Man has climbed up from the Great Mother of the collective unconscious, but he is falling inadvertently into the Great Mother of the affluent society whose nurse is the welfare state"[25]—whose wet nurse, she might also have said, is the compulsory school system. The point is that while, in de Castillejo's terms, "the menace to civilization today seems to come from man's overdeveloped thinking," the very institution that exalts rationality on the surface but which consists of a "smothering atmosphere" in its structure is, in Illich's words, a "magic womb." This suggests that the external economic and political world constitutes the collective masculine; social welfare programs, and even more explicitly the compulsory school system, embody the collective feminine.

Evelyn Acworth in *The New Matriarchy*, a book on the legal status of women throughout recorded Western history, says bluntly that "It is not fortuitous that so many of the great leaders of the Philanthropic Movement of the [nineteenth] century were women, and at the present time it would not be far-fetched to describe the whole conception of the Welfare State as a matriarchal approach to a problem of social life."[26] And, indeed, civilization arose out of the *abundance* of the feminine agricultural village. The emergence of a ruling class, that both enforced surplus production and expropriated that surplus, has led to a long-standing historical misapprehension that the necessary fruits of natural bounty are *scarce*. Socialism is an essentially feminine impulse that seeks to reclaim a shared abundance. The deeper intellectual and spiritual task is to distinguish this impulse from the masculine utopian slave-work system, and then to graft an *ecological* socialism onto a cooperative (and small-scale) agricultural restoration. This misapprehension—nature as niggardly, peasants as a stupid horse needing to be whipped—comes to full synthetic flower in technological society: chemicals and machines for fields and gardens, compulsory scientific indoctrination for the sons and daughters of the farm.

If the "struggle for consciousness is the perennial struggle of the son to break free from the Great Mother," as de Castillejo insists, then the opposing principles contained within the school system provide the setting for a continuous raising of tension: as new fields of specialization are opened, and as new heights of knowledge are attained, the more the classroom engulfs the young, the more extensive and precise become its boundaries, and the more fully does it becomes, for an ever-lengthening period of time, the primary *world* of our youth, compulsory standardization as normative lifestyle. Computerization now intensifies this engulfment, making it instantaneous, universal, and ubiquitous. (Not in the least competent with computer usage, and willfully stubborn about learning, I believe I am beginning to recognize that this intensified engulfment, with its instantaneous, universal and ubiquitous characteristics, also has paradoxical possibilities for exploratory liberation. However, the very complexity and protean sophistication of the computer mimics life and imitates community, even as it seduces the user into greater commitment and deeper engagement. This obsession flirts with a new kind of spiritual addiction, an electronic séance in which the novitiate begins to lust after the illusive, ersatz mind of God—something all-knowing and everywhere present—via the ouija keyboard and computer screen.)

The wrestling of the sexes through the medium of social institutions is, for all practical purposes, completely out of control. It is as impossible to check the growth of schools, or to reduce them to human scale, as it is to curtail the ideology of economic growth or to reverse the buildup of the nuclear arsenal. Again, one runs into the dense psychic morass of these conditions when Illich says that

> . . . transfer of responsibility from self to institution guarantees social regression, especially once it has been accepted as an obligation. So rebels against Alma Mater often "make it" into her faculty instead of growing into the courage to infect others with their personal teaching and to assume responsibility for the results. This suggests the possibility of a new Oedipus story—Oedipus the Teacher, who "makes" his mother in order to engender children with her. The man addicted to being taught seeks his security in compulsive teaching. The woman who experiences her knowledge as the result of a process wants to reproduce it in others.[27]

Illich has an acid pen, and his brilliance is often mixed with an element of macho sarcasm—as in the passage just quoted. Yet he is forever riding the edge of profound psychoanalytical insight, and one cannot help but sense that if Illich could only throw off a certain haughty posture, his profundity as a thinker and writer would even be greater. As it is, he repeatedly offers deep insights embedded in layers of scorn.

The late Paul Goodman, while not given to sarcasm in his writing, was concerned primarily with education for boys. Yet his remarks in *Compulsory Miseducation* apply as well to girls:

> ... the chief mistake we make is to pay too much direct attention to the "education" of children and adolescents, rather than providing them a worthwhile adult world in which to grow up. In a curious way, the exaggeration of schooling is both a harsh exploitation of the young, regimenting them for the social machine, and a compassionate coddling of them, since mostly they are productively useless and we want them to waste their hours "usefully."[28]

School systems will swell and consolidate so long as the industrial economy expands. We will not have human scale or youthful freedom in our schools until we have ecological respect and socialist understanding in our modes of production, for both developments would indicate a relaxation of compulsory civilization, a breaking-through from utopian to eutopian consciousness. Only then will we be ready for the "integration of the personality" that Evelyn Acworth predicts: the creation of what we might call spiritual androgyny.

IV

Let's unfold some of these concerns from a slightly different perspective. Let us begin, shortly, with a passage from Michael Harrington's *Socialism*—and the reader should understand that I have a great deal of respect for Harrington, both his politics and his scholarship. Yet, in reading several of Harrington's books, I have repeatedly felt uneasy about his utopian bias and his neglect of agrarian or rural issues. (It is probably not incidental that he lived, taught, and wrote in New York City.) But while I have at times felt opposition welling up, I have only once, in the course of reading his works, broken out in laughter. The cause for that can be found in the second chapter ("The Preconditions of the Dream") in his *Socialism*. Let me lead up to the point—a small one, perhaps, but revealing anyway.

Harrington says that "Utopia began in the theology of the Hebrews in the desert and the speculation of the Greeks in the cities," and that the "Western utopian tradition" which profoundly influenced European socialism "begins with the Hebrews and the Greeks." In writing of the ancient Jews, Harrington says:

> A nomadic people who lived in between the two highly developed cultures of Egypt and Babylon, the Israelites made an extraordinary intellectual leap to monotheism, in part because they were so backward. Their life in the desert was, as Max Weber has pointed out, so primitive that they did not even have the tools and the artistic tradition to make an icon of their deity. So when they came into contact with the learning and sophistication of their neighbors, their God was able to skip over the stage of polytheism precisely because he was imageless. Then when the tribes settled down, they acquired a parasitic court, a new business class and an impoverished peasantry and proletariat.[29]

Now the sequence alluded to—settled tribes, parasitic court, impoverished peasantry and proletariat—is, of course, the familiar process of utopian unfolding, the dynamic of civilized class structure and exploitation that we have already examined. But the idea that a group of people, which in our time has produced Freud, Buber, Einstein, and Karl Marx, did not have the tools or artistic tradition to make an "icon of their deity" is hilarious. Harrington gives me a real belly laugh. We need to look elsewhere for explanations of Hebrew monotheism.

Merlin Stone, in *When God Was a Woman*, deals explicitly with the development of monotheistic Judaism. Stone attempts to establish the possibility that the Levites, the priestly class of the Hebrews, were originally a separate group of people known as the Luvians who "lived directly south of the Hittites in the area known as Cilicia, close to the Toros Mountains." Stone also identifies the Luvians with other "patriarchal northerners" who swept into the area of the Fertile Crescent as invaders, beginning about 2300 B.C. The Levites, Stone concludes, imposed a stern patriarchal monotheism—the Fatherhood of God—on a group of indigenous people given originally to the worship of "the Goddess."[30] This provocative idea throws an entirely different light on the "backwardness" of the early Hebrews. It suggests the imposition of patriarchal monotheism and not, as Harrington suggests, an "extraordinary" intellectual leap.

And if Plato's "version of the ideal commonwealth" is, in words Harrington quotes approvingly from Marx, "an Athenian idealization of the Egyptian class system," then the prophetic teaching during the time of the Hebrew monarchy "emerged to summon the people back to the simpler virtues of the desert faith, denouncing the capitulation to riches and privileges that had estranged Israel from its God."[31] Perhaps the hidden element within this prophetic movement was an unconscious and repressed yearning for the freedom of pre-Levite society. Harrington again:

> Out of this essentially conservative desire to recall Israel to the old ways there came a magnificent messianic—and utopian—vision of the future: "Then the wolf will lodge with the lamb, and the leopard will lie down with the kid; and calf and the lion will graze together." Thus the idealized past became the future, the simplicity of the nomadic years in the wilderness inspired the description of the golden age to come. After the time of the prophets there was a long silence, yet the utopian spirit they had articulated lived on . . . . And that was the aspect of the Jewish tradition that prepared the way for Christ.[32]

We can interpret that messianic vision thusly: unable to formulate their *actual* attitudes toward the past because of the sternness of their religious system, the Hebrew prophets projected into the future the decisive elements of their violated and imprisoned history. (I am obliged to quote Harrington's words verbatim. Yet the reader should understand that "utopian" in the passage quoted above, regarding lambs and leopards, should really be "eutopian." We are constantly having to cope with the utopian impulse that wishes to dominate the entire analysis. Getting loudmouthed utopia to shut up and shy eutopia to speak is no easy task, even for a writer alone at his typewriter.)

Here is another pertinent remark from Harrington's *Socialism*, a remark that begins to show how Christian imagery began early to take on utopian coloration:

> The ambiguity in all of this can best been seen in Saint Augustine's *City of God*. On the one hand, it is a deeply pessimistic book, reflecting the imminent breakdown of the Roman Empire. The faithful are summoned to turn their back upon all the secular cities—Cain, Augustine recalls, was a builder of cities—and to look only to God for salvation. And yet, the faithful were themselves a city—a city of God. Once that idea was introduced into history, there was a religious basis for trying to build that heavenly community on earth.[33]

This last remark is deeply important and therefore worth repeating: once the idea congealed that the Christian faithful *were themselves a city*, a city of God, then "there was a religious basis for trying to build that heavenly community on earth." When Christianity, on the occasion of Constantine's "conversion," fused with the late Roman Empire, its identity became firmly civilized and explicitly utopian; but that does not mean that either the Jewish prophetic tradition, the Gospels' "kingdom of God," or "primitive" Christianity was utopian. Just as the eutopian Hebrew prophets had spoken out against the theocratic state—both its exploitation and its decadence—so there may also be a way of understanding the unfolding of Christian spirituality that is eutopian rather than utopian. I will deal with this possibility in the next essay. But now we can take a related step in this intellectual unfolding.

V

It is both useful and necessary to deal again with that knotty, historic conflict between Marxism and the "utopian socialists"—that is to say, the conflict between utopian Marxism and the eutopian anarchists. To speak for the latter I bring forward Martin Buber, especially his book *Paths in Utopia*. For the former, I will draw on Michael Harrington's *Socialism*. (Once again, I am obliged to point out that Buber's use of "utopia" is mistaken. As a cooperative anarchist, Buber's path is in *eutopia*.) But I would like to preface my specific remarks on the conflict between utopian Marxism and eutopian anarchism with some references to another work by Buber, *Two Types of Faith*. In the Foreward to this latter book, Buber says "There are two, and in the end only two, types of faith." One begins with the fact "that I trust someone," the other from the fact that "I acknowledge a thing to be true." For the former, Buber says, "status is the decisive thing," and for the latter "the act." In the former case, one "'finds himself' in the relationship of faith," but in the latter "he is 'converted' to it":

> The man who finds himself in it is primarily the member of a community whose covenant with the Unconditioned includes and determines him within it; the man who is converted to it is primarily an individual, one who has become an isolated individual, and the community arises as the joining together of the converted individuals.[34]

So organic membership—one might say membership by birth—characterizes the community by covenant, while membership by election characterizes the community of the converted. Buber again:

> The first of the two types of faith has its classic example in the early period of early Israel, the people of faith—a community of faith which took its birth as a nation, a nation which took its birth as a community of faith; the second is the early period of Christianity that arose in the decay of ancient settled Israel and the nations and faith-communities of the Ancient East as a new formation, from the death of a great son of Israel and the subsequent belief in his resurrection, a new formation which first, in prospect of the approaching End, intended to replace the decaying nations by the Community of God, and afterwards, in view of the newly-beginning history, to span the new nations by the supernation of the Church, the true Israel.[35]

Let us call this "supernation" Christian civilization or Christian utopia.

Finally, Buber flatly asserts that the dynamics within Christianity—its focus on conversion rather than status—"originate essentially from the Hellenistic atmosphere."[36] So, in some measure, we can pick up where we left off—in the dialectical tension between the Hebrew desert and the Greek city, between an earthy, communal eutopia and an urban, aristocratic utopia. And, utilizing David Bakan's terms "communion" and "agency," it's possible to suggest that we can see in the Hebrew model, with its emphasis on organic membership, a feminine ethos; and in the Christian model, with its emphasis on isolated individuals joined through conversion, a more masculine ethos. Let's dig in again with a rather long passage from Buber's *Paths in Utopia* for a beautiful expression of the eutopian purpose:

> At whatever point we examine the structure of such a society we find the cell-tissue "Society" everywhere, i.e., a living and life-giving collaboration, an essentially autonomous consociation of human beings, shaping and re-shaping itself from within. Society is naturally composed not of disparate individuals but of associative units and the associations between them. Under capitalist economy and the State peculiar to it the constitution of society was being continually hollowed out, so that the modern individualizing process finished up as a process of atomization. At the same time the old organic forms retained their outer stability, for the most part, but they became hollow in sense and spirit—a tissue of decay. Not

> merely what we generally call the masses but the whole of society is in essence amorphous, unarticulated, poor in structure. Neither do those associations help which spring from the meeting of economic or spiritual interests—the strongest of which is the party: what there is of human intercourse in them is no longer a living thing, and the compensation for the lost community-forms we seek in them can be found in none. In the face of all this, which makes "society" a contradiction in terms, the "utopian" socialists have aspired more and more to a restructuring of society; not, as the Marxist critic thinks, in any romantic attempt to revive the stages of development that are over and done with, but rather in alliance with the decentralist counter-tendencies which can be perceived underlying all economic and social evolution, and in alliance with something that is slowly evolving in the human soul: the most intimate of all resistances—resistance to mass or collective loneliness.[37]

Just as a "city of God" and a "supernation of the Church" are explicit utopian constructs that silently announce the obliteration and defeat of eutopian spirituality, so we also can see the "hollowing out" of society by capitalist economic techniques as a secular but parallel process, resulting in "the whole of society" becoming "in essence amorphous, unarticulated, poor in structure."

In the first decade of the twenty-first century, the Left is full of hand wringing over its collective powerlessness, its inability to engage substantial issues or galvanize a meaningful electoral majority. Hardly anywhere is there an adequate grasp of how thoroughly utopia has penetrated society and reshaped human consciousness, of how mass-produced utopian concepts have replaced the folk consciousness of what Buber calls "the old organic forms." Our grasp of this loss is anesthetized by an induced disgust for the past with its "backwardness" and "scarcity," a disgust for the folk and agrarian past whose flip side is a kind of religious craving for "progress," for the new and the novel, for the fulfillment sure to come with the newest utopian discovery or invention.

As the magnitude of our crises intensifies, as utopian consciousness becomes increasingly incompetent, bizarre and crazy, eutopian consciousness remains in short supply. And so the political will to engage these crises in meaningful ways is lacking precisely because political will is so fully blinkered and channeled by utopian fantasy and expectation. The Right is in full demented grip of this utopian mythology. The Left is in full-spectrum hangover, with many wanting new forms of the old drug, and some startlingly sober few beginning to grasp the horrendous immensity of the utopian disaster building in intensity. Between these two there is a multitude with acute symptoms of nausea, headache, bewilderment, and shortness of breath, uncertain whether to belly back up to the bar or to renounce once and for all the utopian drug. And what, exactly, would it mean to renounce the utopian drug? What *are* the politics of eutopia?

In his *Socialism*, Michael Harrington insists that the "utopian socialists"—he names Robert Owen and Claude Henri di Saint-Simon explicitly—attempted to create eutopias under the umbrella of capitalism; that these eutopians "wanted to withdraw totally from a corrupt system,"[38] implying that by their withdrawing the "corrupt system" of capitalism would remain unchallenged and continue essentially intact. Harrington suggests that eutopianism is a form of political quietism, in other words, based in passive retreat. Harrington does state, although he fails to elaborate on the idea to any significant extent, that a "vision of fundamental transformation of social relationships" is at the heart of socialism. He also says that, in the context of nineteenth-century industrial England, Marx was "the first thinker to realize what was happening and to make it explicit; he was the consciousness of the dream, the moment when it awoke to itself."[39] This is subtle thinking—subtle more in what it doesn't say than in what it does. Socialists for the most part (and this includes Harrington) are rigorously disinclined to indulge in what they derisively call "utopian speculation." But once one gets behind the semantic veil, the reason for this disinclination is apparent: mainstream socialism is itself a highly-charged utopian ideology that, except for the elusive eutopian notion of the "withered state," rationalizes the regimentation of utopian perfection in the name of Progress. This ideology requires an imposed and artificial social structure to maintain economic growth and develop a "civilized" standard of living—exactly the kind of society that Buber calls "amorphous" and "unarticulated," "hollow in sense and spirit—a tissue of decay."[I]

In the second chapter of his *Paths*, Buber once again creates a twofold distinction. On the one hand, the "vision of rightness in Revelation is realized in the picture of a perfect time—as messianic eschatology"; and, on the other,

I. In his Preface (1962) to the reprint of *The Story of Utopias* (1922), Lewis Mumford says on page 5 "Though Marx and Engels mocked at utopian socialism, their followers, for lack of any better precedent, smuggled many utopian social inventions through the back door."

> . . . the vision of rightness in the Ideal is realized in the picture of a perfect space—as Utopia. The first necessarily goes beyond the social and borders on the creational, the cosmic; the second necessarily remains bounded by the circumference of society, even if the picture it presents sometimes implies an inner transformation of man. Eschatology means perfection of creation; Utopia the unfolding of the possibilities, latent in mankind's communal life, of a "right" order.[40]

(Once again, I ask the reader to read "Eutopia" where Buber says "Utopia," and to infer "Utopia" where he alludes to "perfect time," "messianic eschatology," and, soon to come, "apocalyptic eschatology.") Buber insists that there are "two basic forms of eschatology: the prophetic . . . and the apocalyptic." The first is characterized as that which "at any given moment sees every person addressed by it as endowed, in a degree not to be determined beforehand, with the power to participate by his decision and deeds in the preparing of Redemption." The second is characterized by that "in which the redemptive process in all its details, its very hour and course, has been fixed from everlasting, and for whose accomplishment human beings are only used as tools, though what is immutably fixed may yet be 'unveiled' to them, revealed, and they be assigned their function":

> The first of these forms derives from Israel, the second from ancient Persia . . . . In the socialist secularization of eschatology they work out separately: the prophetic form in some of the systems of the so-called Utopians, the apocalyptic one above all in Marxism (which is not to say that no prophetic element is operative here—it has only been overpowered by the apocalyptic).[41]

Marxism is, clearly, a form of utopia. But if the Marxist utopia is based on apocalyptic revolution, it should also be clear that Christian fundamentalism—the "conversion" principle at fever pitch—is saturated with apocalyptic imagery and is therefore an obvious form of religious utopia, the "redemptive process . . . fixed from everlasting." (It is not meaningless that many of the "neocons" who surrounded the George W. Bush administration and who were its intellectual cheerleaders were, in their youth, Marxist ideologues craving what Buber calls a "messianic eschatology." Though many of them may privately be atheists, their alignment with the Armageddon-bound Christian Right is based on an underlying fusion of utopian lust for something that exceeds and surpasses all earthy creatureliness. All such creatureliness is backward, stupid, limited, pagan and pernicious, and is only deserving of utopian contempt. Therefore many "neocons"—though some may have come along for reasons of sheer predatory greed, particularly in regard to Iraq's oil deposits—are really utopian idealists whose stunning fantasies of global regime change are amazingly categorized as "conservative"! That these varieties of ungrounded fantasies and fantasts are unhesitatingly provided such categorization indicates how weirdly normative utopia has become in the contemporary world.)

Eutopia, then, concerned with "perfect space," orients itself toward small and interdependent communities, integrated by networks of confederation, whose spiritual and cultural attainments are to be realized and enhanced in the here and now. The Marxist utopia, on the other hand, according to Buber's analysis, concerns itself with "perfect time"—as in Marx's objection to the eutopians that the time for revolution was not yet ripe—and with planning toward the ultimate revolution. The eutopians see utopian Marxists as destroying the fabric of already existing communal forms in favor some unspecified future perfection, theoretical in nature; the utopian Marxists claim that the eutopians, by wandering back to the village hearths, dampen the ardor for Redemption and suffocate the desire for revolutionary civilization. As Buber says, "The Utopia of the so-called utopians is pre-evolutionary, the Marxist one post-revolutionary."[42]

Eutopians hold that fulfillment is the real point, the real goal: technical advance ("progress") is to be controlled within and by the social body, relegated to its proper place, and not allowed to destroy the social tissue. (Eutopians in utopian society therefore call for reform by select retrogressions. And, since Green politics calls for an "ecological society," which can only be created and sustained by the implementation of select retrogressions, Green thinking is necessarily eutopian.) The "destination" of the eutopians, in other words, is here and now; and the realization of the here and now requires the throwing off of consuming preoccupations with the future, with "progress," and a building onto or building up of those communal forms that already exist or have managed to survive in utopian society. Utopians contend that all remaining communal patterns are unsatisfactory, that they are tainted by hidden self-interests (including sexual stratification) or are simply regressive and "backward." Eutopians fear that the atomization of society, the grinding to pieces of every last vestige of organic community (of which kind of community small-farm neighborhoods are, or were, a pre-eminent example), only portends an increasingly bureaucratic and technocratic regimentation, and that the utopian expectation of real community *at the end* of this atomizing process is at best wishful thinking and at worst an ideological smokescreen for yet another dimension of totalitarian utopianism and even ecological mayhem.

The reader will easily have seen that my bias throughout these essays has been consistently toward fulfillment and eutopia. But does that mean utopia is empty of meaning or fruitless? There are at least two aspects of utopia that seem to justify its attendant carnage—if one might dare use the word "justify" in this context. These aspects are a global world-view and a new dispensation between the sexes: the keys to the transformation of civilization and the reaching of a fuller human consciousness. It may be that only in our own time, as the utopian-civilized system has reached both its limit to expansion and its approach to breakdown, that we can finally grasp utopia's hidden purpose. Its unconscious goal all along has been eutopian, a eutopia in which there is a fundamental reconciliation between the sexes, a eutopia in which the masculine ethos of domination, fuelled to unprecedented deadly extension by the industrial revolution, gives way to a more feminine ethos of nurture, ecology, and community. As the technology of power dabbles with disaster, Green politics rises in protest from the utopian gridwork of atomization.

"Unless a seed falls into the ground and dies," said Jesus, "it remains alone; but if it dies, it brings forth much fruit." The eutopians have, in a sense, been these seeds, germinating in the dark; the moment of their bursting forth may very nearly be at hand. Marx, it seems, arrives haggard in a rumpled suit, carrying an unassuming suitcase containing only the most meager of personal belonging: a change of clothing, a comb, and a toothbrush—*not* a blueprint for eutopia, for it apparently was his faith that eutopia would take care of itself once the proper "preconditions" were fulfilled. It was the faith of a genius who knew the limits of his thought—Karl Marx, the Moses of Eutopia.

VI

But is Marx the Moses of Eutopia? Or is he yet another member of the fraternity of arid rationalists who avoids the deeper implications of his own thought? An argument can be made for the latter view. Let us look at Marx again, briefly, through the eyes of Walter Weisskopf. Weisskopf's book, *The Psychology of Economics*, was published in 1955—the same year as Herbert Marcuse's *Eros and Civilization* and four years before Norman

O. Brown's *Life Against Death*, books that opened new dimensions in psychoanalytic thinking. In zeroing in on Marx's development of the labor theory of value—a concept earlier expounded particularly by the classical economist Ricardo—Weisskopf makes some important observations.

Given that labor, already in the early decades of the nineteenth century, is perceived as primarily masculine in nature, the labor theory of value tended, in the words of Weisskopf, to "emphasize the importance of the male element in the creation of economic value . . . to the exclusion of the female part."[43] And this "female part," especially in the theories of Malthus, was identified as the "source of all economic ills, the evil in [his] universe of discourse." Malthus expounded a theory for which he is famous to this day: namely, that overpopulation and underproduction are "natural laws" attributable to the "inferiority" of the female. On the one hand, overpopulation is the result of "irrepressible fecundity" in the human female; on the other, underproduction of food is the result of "nature's niggardliness." Nature generally, and the Earth specifically, are identified by Weisskopf as symbolically "female"; therefore, the disparagement of the natural productivity of the Earth is perceived by him as an unconscious disparagement of both "female" nature and female human beings.

In this context, the labor theory of value—the idea that *only* labor, particularly *male* labor, creates value—is picked up by Marx and refined. Writing of Marx, Weisskopf says that

> Instead of attributing the special quality of surplus productivity to land, he imputes it to human labour power. The mystery, the secret of productivity, the faculty of being able to produce more than there was before, to create something new, is now attributed to labour. This means that "fertility" in respect to value is transferred from the "female" land to the "male" factor, labour.[44]

Perhaps it is fair to say, psychoanalytically, that the labor theory of value is an intellectually sublimated form of womb envy, another urban victory over fertile soil. Weisskopf shows that Engels had picked up on the works of the early American anthropologist Lewis Henry Morgan as well as the studies of the Swiss interpreter of ancient myth Johann Jakob Bachofen; Engels postulated a prepatriarchal period of primitive matriarchal communism. Weisskopf quotes Engels: "The defeat of matriarchy represents the defeat of the female sex in world history." Weisskopf goes on:

> ... Engels, in his analysis of the bourgeois family and society, takes the part of the woman against the male head of the family. She is the exploited part, the "proletariat" in the family; her position is represented as similar to the one of the wage labourer in capitalist society. Engels establishes a close interrelation between male predominance, private property, and capital, the main targets of his criticism. He is also very intent on proving that real love and sexual passion cannot flourish in bourgeois marriage and have to be liberated by the socialist revolution.[45]

Given these thoughts as background data to be interpreted in psychoanalytic terms, Weisskopf says that

> Thus we arrive at two pairs of antinomies which can be co-ordinated with the thought of Malthus and Ricardo, on the one hand, and of Marx and Engels, on the other: capitalist versus anti-capitalist, and anti-female versus pro-female. This suggests an interconnection between those two pairs of opposites. The apologists are anti-female, the antagonists are pro-female. At the same time, the antagonists are anti-male, at least as far as the father, the head of the bourgeois family, is concerned.[46]

The twist or reversal in the "pro-female" and anticapitalist stance of Marx and Engels lies in the labor theory of value. Marx, picking up the labor theory and installing it as a cornerstone in his industrial analysis, pushes aside the "pro-female" element in his revolutionary consciousness and restores the principle of masculine power. Weisskopf sees this dynamic in terms of the Oedipus complex: the capitalist system is the "father" and the revolutionary "sons" set themselves against "him," using in their struggle, however, a major conceptual device of the capitalist "father" they seek to overthrow: the labor theory of value.

Earlier in this essay, I invoked David Bakan's dialectical terms "agency" and "communion." In the second chapter of *The Duality of Human Existence*, Bakan translates these concepts into a political context:

> The Protestant ethic is allied to the spirit of capitalism in that both entail the vaulting exaggeration of the agentic feature of the psyche and the repression of the communion component. Later in this essay, I deal with the matter of sex differences and point out that, although the agency and the communion features are present in both males and females, agency is greater in males and communion greater in females.[47]

This leads to the conclusion that the labor theory of value is an example of "the vaulting exaggeration of the agentic feature." The eutopian revolution is avoided because, in the final analysis, the "sons" seek to step into the "father's" utopian shoes. The "female" elements—practical equality for women on the one hand, ecological respect for the Earth and political respect for rural culture on the other—are given low priority because they interfere with maximized industrial productivity, with unlimited human control over nature. The Marxist revolution founders on its own theoretical flaws—its uncritical enthusiasm for civilization and its debilitating male utopianism.

Indeed, the emergence of civilization from the abundance of the agrarian village is, for all practical purposes, defined by an elite forcibly concentrating this abundance in the hands of a coercive minority. Civilization was created by this theft both of the abundance itself and of the labor that facilitated this abundance, as well as by the expropriation of land. Capitalism, whether private or state, is the economic secularization of this centralizing principle within utopian civilization. By forcibly "stepping into the father's shoes," revolutionary Marxists entered into an ideological (and spiritual) contradiction. They were in principle for labor, which means equality (roughly speaking) and economic decentralization in the direction of eutopia. But in practice they concentrated power—Stalin is the poster child—in a massively intensified effort to build a capital-intensive industrial infrastructure precisely at the expense of the peasantry, small farmers, and agrarian culture.

In this respect, the Stalinist concentration was an exact replica of the capital formation process within the earliest civilizations; and, for all the pious talk about "labor theory of value," the kind of Marxism practiced in the twentieth century, most certainly in its Soviet form, was as rabidly civilized and utopian as any kind of civilized economy in history. Civilization is inherently capitalist and is inherently built on coercive theft. This is the conundrum that all revolutionaries of the Left are faced with: a vision of decentralized liberty, of ecological stewardship and humanitarian servanthood, but with an inherited mythology and methodology of coercive power—an unresolved chasm between eutopian wholeness and utopian compulsion.

VII

So how can these opposing principles of utopia and eutopia possibly be synthesized? If we start from a position that identifies conflicting principles in cultural history as polarities within a single process of transformation and development, we are then, I believe, on the right path. This does not, in and of itself, suggest what our ultimate destination is (it does suggest a synthesis but not the ultimate content of that synthesis), but it provides us with an image that does suggest that the conflicting tendencies within utopia and eutopia can and will be resolved. A synthesizing image comes from Johann Wolfgang von Goethe's conception of the growth of plants. If one visualizes a tree and considers the trunk, the upward thrust, that which reaches up in time, as masculine, and both the branches and roots that reach respectively into space and earth as feminine—then one has a useful and substantial image for a synthesis of opposing forces, neither of which could exist without the other.

In Peter Salm's book, *The Poem as Plant*, one finds an attempt to integrate the conceptions of Goethe's nonmanipulative science, explicitly his studies on the morphology of plants and color theory, with the constructions of his art, *Faust* in particular. Salm is after an "organic unity," and he finds it in Goethe's conception of the "archetypal plant." The growth of this plant, Salm writes,

> . . . takes place as an interplay between opposing forces—the "vertical tendency" and the "spiral tendency." The former is that force which urges the plant in a "straight line toward the heavens"; the latter is the nourishing, multiplying principle which gives solidity to the plant. Only when both tendencies are in a dynamic balance, and in harmony with the general environment, can the plant develop normally without becoming a monstrosity. In concluding his late essay *Spiraltendenz der Vegetation*, Goethe associated these opposing tendencies with the male and female principle: ". . . if we now see how the vertical tendency proves itself to be decisively masculine and the spiral tendency decisively feminine: then we can view all vegetation from the root up as though in secret and androgynous union; whereupon, as a consequence of the transformations inherent in growth, the two systems separate in clear contrast and juxtaposition, only to be united again in a higher sense."[48]

So here we have the "vertical tendency" and the "spiral tendency," the straight line and the nourishing principle, the two systems that separate only to be united.

Are we ready now, historically and politically, to slow the vertical tendency and consciously facilitate the spiral tendency: to radically reduce energy production and consumption, decentralize political power, promote the reconstruction of rural life and culture? To rein in utopia and resurrect eutopia? How else can one reconcile what Harrington, in *Socialism*, says are Marx's "two practical preconditions" for socialism—abundance and alienation—with what Buber aptly calls the "modern individualizing process finishing up as a process of atomization"?[II] Buber rightly throws that question in the face of utopian Marxists; yet, it may be that only in the modern context of atomized urbanization, social conditions generated by the industrial utopia, that women could have gathered sufficient power—precisely because of technological alienation!—to force the issue of their "second sex" into the political arena and demand social, economic, political, and, yes, religious redress.

The symbol of the tree tells us something else: in reaching the limit of its upward thrust (as the utopian system is now reaching its upward limit), the primary function of the tree passes to the branches and into the production of organic fruit. The mature tree, solid of trunk and spacious of bough, throws its magnificent energy into the creation of androgynous community. As Salm writes:

> The plant progresses by expansion and contraction, and the moment of flowering brings with it a diminution by metamorphosis of the basic leaf to form the various parts in the flower. The life force and upward thrust of the plant in bloom have been suspended in favor of concentration and refinement, to prepare for the plant's highest goal, sexual propagation. Goethe uses the botanical term

II. Harrington says on page 30 that "The second precondition came about because the very development of those means of production 'had made the mass of mankind "propertyless"', and since the might of its own social productivity loomed over it like an alien power, revolutionary." Harrington is here quoting Marx directly, and it is significant that being "propertyless" (having, as biological beings, no right or entitlement to land, to Earth) is one of the definitions of "alienation." Harrington specifies that this second precondition requires a substantial body of women and men who have overcome alienation and who stand ready to redirect the means of production in the service of humane democratic socialism. But it is also instructive to note, in light of the discussion on the labor theory of value, that it is *productivity* that looms over the masses "like an alien power, revolutionary." Indeed. It is the "father's" ghost.

> "anastomosis" for this state of concentration and refinement, the interruption of the vital upward thrust, a kind of sleep or symbolic death. Space contracts into a single point at which length and breadth are annulled.[49]

This is like the tree of which Jesus spoke, a tree that begins as "the least of all seeds" and culminates in a massive and splendid growth in which the birds of the air can come and make their nests—natural imagery, eutopian.

Natural imagery based on successive stages of growth and development can also be used to describe cultural change: a precivilized period of the Mother; a stern and rational civilization of the Father; a democratic-industrial civilization of the Son; and a postcivilized era of the Daughter. These stages or "eras" can be seen as cultural and spiritual progressions, an evolutionary spiral whose transitional points might be more aptly characterized as revolutionary. Given this imagery, it's possible to say that we are now at the edge—in a very real and immediate sense—of the Daughter's era. The era of the Mother gave way long ago; although Evelyn Acworth, in *The New Matriarchy*, says that the

> . . . transition from matriarchy to patriarchy was one which took place over a long period of time. During the early historic period, races and people existed in various stage of transition, often showing an admixture of matriarchal and patriarchal customs or beliefs, their development depending on external circumstances and also on the degree to which warlike characteristics had been developed.[50]

The era of the Father has not yet passed away—as is evidenced, for example, by the political and religious strife between rival Father systems in the Middle East and Asia. Both the capitalist West and the communist East are manifestations of the Son's era, and their strife is a fraternal one—although "fraternal" in this sense doesn't make the potential consequences of their strife any less catastrophic. If anything, the consequences of modern warfare point to annihilation. Norman O. Brown, in *Love's Body*, puts the predicament succinctly: "The old agonal warfare was between brothers; conducted according to rules; limited in objectives, and limited in time, in a necessary alternation of peace and war; the brothers need each other in order to fight again another day. The new warfare is total: it seeks an end to war, and end to brotherhood."[51] Considered psychoanalytically—strife between Fathers, strife between Sons, strife between Fathers and Sons—the predictions of Christian fundamentalists regarding a Middle

Eastern Armageddon are not so farfetched after all. It is in the area of these (oil-producing) countries where two of the religions of the Father, and the patriarchal social structures that derive from these religions, are centered. Perhaps one might even say that, in a biblical sense, such a war would be a war of fossil inheritance with fraternal strife in shifting alliances among the heirs. In that instance, the cradle of civilization would transmogrify into civilization's coffin.

If war is the "special invention of civilization," as Lewis Mumford has asserted in *The Transformations of Man*, its "ultimate drama," then we might well expect that "If organized force brings civilizations into existence, organized force is likewise the agent of disintegration that ultimately brings them to an end."[52] And men can be expected to discharge their fearful energies upon each other in that "ultimate drama" until the very nature of their fraternal identities is transformed—and, with that transformation, the discovery of a spirituality far more embracing of the feminine.

Can we dare to say that the Father will lie crushed, the Son exhausted, and that the Daughter will step forward with unanticipated political strength to transform the prevailing social ethos? This is not to say that we will have instantaneous eutopia or that new and even greater crises do not await our descendants. Nor am I saying that men are evil and women good. Women, too, when they begin to rule and dominate—and they will—will reach a dead end; and they will do so even more quickly than men have, because of technology. Women will discover, to their dismay and chagrin, that hubris is not the exclusive preserve of men. In her book *The Creation of Patriarchy*, Gerda Lerner says, in fact, that

> Perhaps the greatest challenge to thinking women is the challenge to move from the desire for safety and approval to the most "unfeminine" quality of all—that of intellectual arrogance, the supreme hubris which asserts to itself the right to reorder the world. The hubris of the godmakers, the hubris of the male system-builders.[53]

The handwriting, as it were, is already on the wall.

In the meantime, but after the coming breakdown, political power will be shared; but first the barrier to a fuller world, the institutions of male hegemony, will be broken. Then the work of reorienting the social fabric in a eutopian dimension will begin. We will, at last and for the foreseeable future, be on the path of balance and integration: reconciliation between sexes and races, an abundance of international communication and travel, the recovery of bioregional cultural coherence. Why is breakdown necessary as a prelude to eutopian reformation? It is *the breaking down* of barriers and obstacles that is at issue, and these obstacles and barriers run the full spectrum from the personal and subjective to the institutional and objective. They include personal world-view and habit, religious mythology, national identity, and institutional prerogative. It's true that—theoretically—the transformation could be achieved peacefully, by means of rational policy and political will.

But peaceful change can be achieved only if we overcome our ideological and mythological fixations; only then could the brutalities which are increasingly the only alternative be avoided. Either repentance with hard choices and spiritual suffering or avoidance that leads irrevocably to disaster. Yet we have, collectively, been so deadened to the complexity of issues and so misinformed as to their content, that "peaceful change" at this point means nothing less than a thorough re-education that deepens into painful identity crisis and spiritual rebirth. And *that,* given the powerful forces within our inherited religious and ideological fixations, may be more than we can rationally expect. History seems to indicate that rational persuasion without the pressure of powerful external events is of limited applicability. The "haves" invariably disdain rational exercise in self-examination when the question of their personal intellectual and spiritual flabbiness is the issue at hand. We are all too pleased with our "self-image" and "self-esteem" to entertain the desire of disturbing that imaginary reflection by a deep and searching look into the mirror which the world's powerless, poor and dispossessed—the "have-nots"—hold hopefully up to us. Only by looking into that mirror, and letting our shame burn the illusions from our hearts and minds, will we be able to avoid continuing civilization's "ultimate dramas."

The world is so large and so complex that no organization, no person, no call for peaceful change can really expect—although we are all unconditionally required to continue our efforts—to check the overall direction toward breakdown. I do not wish to be misunderstood: one *must* take a responsible and active part in the conflicts and issues of life, both public and private, toward the desired end of peaceful change. Only those who do so can dare to look over the rim of our present world. Those who nurse private desires and indulge in utopian ideologies will find the utility of their wisdom commensurate with their horizons.

Yet history proceeds inexorably. If we as individuals and as whole societies are not willing to fulfill our destinies voluntarily, we will have then thrust upon us, irrespective of our "permission" or consent, and with considerably less grace than if we had undertaken them ourselves, responsibly. And shall we then, with all the warnings offered us repeatedly, pleadingly, plead ignorance of the suffering in the backyards of the world and attempt to swear off any and all personal and social responsibility?

We will be shaken to our roots, the final convulsions of our age. Exhausted by war, weary of ourselves, we will be both forced and freed to grow beyond our old selves. But we still have a loftier—and an earthier—path to travel. For those who are able to discern the future, the transformation is already in their midst. The harvest is heavy, but the laborers are few.

Creator Spirit come.

NOTES

1. Illich, *Deschooling*, 105.
2. Illich, *Deschooling*, 105–6.
3. Illich, *Deschooling*, 106.
4. Illich, *Deschooling*, 107.
5. Illich, *Deschooling*, 115–16.
6. Freud, *Civilization*, 154–55.
7. Castillejo, *Knowing*, 47.
8. Lerner, *Creation*, 40.
9. Lerner, *Creation*, 44–45.
10. Mailer, *Prisoner*, 116–17, 111.
11. Castillejo, *Knowing*, 105.
12. Castillejo, *Knowing*, 39.
13. Bakan, *Duality*, 14–15.
14. Bakan, *Duality*, 110.
15. Castillejo, *Knowing*, 50.
16. Castillejo, *Knowing*, 42.
17. Illich, *Deschooling*, 32.
18. Castillejo, *Knowing*, 45–46.
19. Aries, *Centuries*, 284–5.
20. Mailer, *Prisoner*, 127.
21. Castillejo, *Knowing*, 48–49.
22. Castillejo, *Knowing*, 53.
23. Castillejo, *Knowing*, 49.
24. Castillejo, *Knowing*, 50–51.
25. Castillejo, *Knowing*, 48.
26. Acworth, *New*, 159.
27. Illich, *Deschooling*, 39.
28. Goodman, *Compulsory*, 183.

29. Harrington, *Socialism*, 13–14.
30. Stone, *When*, 100.
31. Harrington, *Socialism*, 15, 14.
32. Harrington, *Socialism*, 14.
33. Harrington, *Socialism*, 15.
34. Buber, *Two*, 7–9.
35. Buber, *Two*, 9–10.
36. Buber, *Two*, 11.
37. Buber, *Paths*, 14.
38. Harrington, *Socialism*, 27–29, 38.
39. Harrington, *Socialism*, 9, 29.
40. Buber, *Paths*, 8.
41. Buber, *Paths*, 10.
42. Buber, *Paths*, 10.
43. Weisskopf, *Psychology*, 146.
44. Weisskopf, *Psychology*, 151.
45. Weisskopf, *Psychology*, 131.
46. Weisskopf, *Psychology*, 131.
47. Bakan, *Duality*, 20.
48. Salm, *Poem*, 27.
49. Salm, *Poem*, 107.
50. Acworth, *New*, 33.
51. Brown, *Love's*, 19.
52. Mumford, *Transformations*, 47.
53. Lerner, *Creation*, 228.

26

Keeping the Lid on Jesus' Coffin

In the previous essay, I said I would deal with the possibility of eutopian unfolding within Christian spirituality. On the one hand, this is easy to do, especially with such a lucid scholar as Marcus Borg to rely on. (I am thinking here, particularly, of the sixth chapter in *The God We Never Knew*, entitled "The Dream of God: A Politics of Compassion.") As Borg says, some ways of thinking about God legitimate the existing political order, and other ways lead to a passion for social transformation.

Arriving too late to provide little more than a provocative epigraph quotation for this book as a whole, John Dominic Crossan's *God and Empire* nevertheless confirms many of the assertions made in these essays, as well as the specific "eutopian unfolding" thrust of this essay in particular. Crossan insists that the primary proclamation of Jesus is the kingdom of God and that this proclamation is totally nonviolent, one hundred percent spiritual *and* one hundred percent political, and "it is not that we are waiting for God, but that God is waiting for us."[1] But I am no theologian. Therefore I will try to limit my remarks to what I know, or at least what I think I know.

At one level, I would even suggest that to arrive at a eutopian understanding of the Christian faith it is nearly irrelevant how the "believer" thinks about God. That is, God-talk or God-image may be really quite peripheral as to whether, as a self-identified Christian, one takes seriously the teachings and actions of Jesus as a doorway into a fuller life, or whether one piddles with those teachings and actions, carefully sandbagging them (perhaps with any number of biblical passages or ecclesiastical creeds) in order to minimize their claim on one's life or to shrink their social implications. But if one takes them seriously, it is simply impossible to read the Gospels and not be stunned and either terrified or exhilarated (or terrified *and* exhilarated) by the vision, boldness, courage, and extraordinary confidence of the man whose life-legend lives on. This is not mere hero worship. This is recognition of an exceptionally clear and decisive articulation of how human life could be lived on Earth. To unfold this articulation is complex.

First, I would agree with Borg that the monarchical model of God has served to legitimize the existing political order and, furthermore, that afterlife salvation has been the carrot and hell has been the stick that helps to perpetuate the monarchical model. God the King is also God the Judge. Afterlife salvationism was the slide by which Christianity ended up in the aristocratic utopian camp. This tendency occurred very early. Apparently anticipating the imminent end of the world, the author of Romans 13, in the Christian *New Testament*, admonishes (verse 1) "every person" to "render obedience to the governing authorities, for there is no authority except from God, and those in authority are divinely constituted." Such an assertion implicitly contains both a repudiation of the kingdom of God as a social vision to be struggled for—that is, an achievable social program of radical servanthood and radical stewardship—and the substitution of explicit otherworldly salvation for such a struggle. It is this repudiation/substitution that Augustine, several centuries later, uses to justify and exonerate the merging of church and empire in Rome, and it is largely this utopian substitution that Michael Harrington believed was the "religious basis for trying to build that heavenly community on earth," an impulse, with a mostly unconscious eutopian core, attempting to manifest itself via utopian means.

With Constantine, the Christian structure had fallen under the rulership of an elite explicitly predisposed to an aristocratic world-view, and this world-view found an easy alliance with the monarchical image of God. They were two peas in a pod. They supported and reinforced each other's agenda. And they revitalized what Crossan calls the "brutal normalcy of civilization"—revitalized it and removed the radical critique contained within the eutopian kingdom of God articulation. Overlay this configuration onto the peasant (i.e., "pagan") base of society and it's easy to see how the explosively visionary Jesus was largely kept in an ecclesiastical box (or hanging on a cross, as a mute but deadly object lesson) through centuries of tight social control—a control in which afterlife hope of heaven and dread of hell played an immeasurable but crucial part in keeping the lid on Jesus' coffin.

The peasant life, the peasant village, overawed by elite extraction and aristocratic control, was precisely the social milieu in which Jesus moved. Etymologically, as a person from the countryside, Jesus was also a pagan. His message was yeast for peasant foment. (For all my admiration of Professor Crossan's scholarship, I think he misses something crucial in his unabashed admiration for the Apostle Paul. That is, he clearly recognizes that Paul focuses on Roman provincial capitals in his proselytizing—"Paul is a man of capitals"[2]—but Crossan uses the word "pagan" in a typical and conventional way, essentially equating "Gentile" with "pagan." Paul, with his urban orientation, is already abandoning the peasant revolutionary implications of the kingdom of God.)

The Benedictines broke off early (sixth century) into community; but, for all their great contributions, they were a celibate monastic order. Their community explicitly prohibited the messy yeast of sexual reproduction, of bawling kids, of a spiritually intentional life inclusive of families—as if the "dream of God" was only for religious celibates. (The Benedictines may have been the unconscious model for Michael Harrington when he blasted "utopian socialists" for withdrawing from a "corrupt system." The celibate orders, at their best, represent the kingdom of God in a prolonged holding action and survival mode.)

For a fuller social inclusiveness, we have to wait until the Protestant Reformation, and then for the persecuted reformation within the Reformation—namely the so-called Anabaptists, of whom, in this country, the Mennonites, Amish, and Hutterites are the most well-known. Here, at last, are ordinary peasant people attempting to live out the kingdom of God in a cooperative and communal manner. But we need to remember—as an illustration (if we need one) of how cruel a grip the monarchical model of God can hold on the human mind—that Catholics, Lutherans, and Calvinists openly and publicly *persecuted and killed* the "Anabaptists" who defied obedience to the state, the prince, the pope, and the state church, and who refused to bear arms or take a loyalty oath. A monarchical God is the model for utopian Christianity. "Heretics" have paid dearly for their resistance to the monarchical model of both church and state. That's not to say that the Anabaptist sects have *achieved* the kingdom of God (or even that their overarching theology is not confounded and compromised by inherited biblical language that emphasizes the monarchical model), but their recognition of the importance of the kingdom of God concept, as a social vision, can hardly be overstated. And it's certainly not accidental that the preacher whose name, in America, is most strongly linked to the "social gospel"—Walter Rauschenbusch—was a Baptist.

The combination of industrialization and the necessary, but overrated, democratic revolutions in the West—"overrated" because we have largely been taught these revolutions were finished products rather than an infant's first wobbly steps—has served to seriously blur and disguise the magnitude of ongoing elite control in both the economy and in politics. (If, as we have seen from Marcus Borg in a previous essay, "elites" and "retainers" held "about two-thirds" of society's "annual production of wealth," with "about half going to the top 1 or 2 percent" in preindustrial and predemocratic social structures dominated by an aristocracy, then it is a very serious warning signal when, according to Kevin Phillips in *American Theocracy*, "By credible calculations, the top 1 percent of Americans in 2000 had as much disposable (after-tax) income as the bottom one hundred million or 35 percent of the population."[3] I submit it is not possible to live democratically in a society with such an aristocrat/peasant wealth differential. I further submit that it is precisely the monarchical image of God, as propounded and promoted by the Christian Right, that allows "democratically" this steady encroachment and enlargement of aristocratic wealth concentration.)

Nationalism is, in turn, largely the means by which the lower classes are lured and coerced into political alignment with those who rule, manipulate, and exploit them. Patriotism, at its root, is about *pater,* father, and right below the waterline of consciousness it merges imperceptibly into the monarchy, the kingliness, of God. "God and country," "God and king," are no idle or accidental slogans. They unite and align us against ourselves. They structure our lives in precisely those ways that perpetuate our consumerist lust for elite trinkets, while extracting "surplus value" from our economic lives and gridworking our educational system with compulsory practice in systemic regimentation. We are addicted to utopia, and the authorities keep us well-drugged with fresh doses of daddy pills. We are so saturated with the normality of utopian deferment (or with its commercial trinkets) that eutopian insights, while instantly attractive and stimulating, are typically repressed as unrealistic and unattainable. God the King is always at hand to sternly rescue us from any lapse into "pagan fantasies."

But eutopian Christianity is what Marcus Borg calls the dream of God. It is what Jesus called the kingdom of God (Luke) or the kingdom of heaven (Matthew). Anyone at any time who has taken the Gospels seriously has been infected with this eutopian, yeasty virus, even in the long and powerful saturation of utopian governing images to the contrary. Eutopia is utopia's living ghost, the never-dead skeleton in the closet, always capable of an unanticipated resurrection. Eutopia is to utopia what the kingdom of God is to otherworldly salvation. Eutopia is the unfolding of Spirit, the Age of the Daughter.

There is something almost manichean regarding Jesus' fierce assertion about serving either God or Mammon (the elite system). The vehemence of his opposition to *the system* of elite control indicates two things: the hugely overbearing and seductive power of that system, and the magnitude and urgency of his vision by which he understood how human life could and should be lived on Earth *without* the controlling dictates of that overbearing system. He invited the skeleton out of the closet, fully clothed in healthy flesh. Utopia has now gridworked the entire Earth. This is the meaning of "globalization." The United Sates is, at this moment, in the control room. Population explodes, resources are recklessly consumed, agriculture is utopianized, a large handful of huge corporations dominate the economy of the entire world, the U.S. military intervenes at will, apocalyptic weapons abound, Earth's ecology shreds and sags.

The "almost manichean" demand of Jesus—is it God or Mammon you will serve? Creation or utopia?—is precisely the question, the demand, that constitutes our existential moment. Earnest religious scholars like Marcus Borg and John Dominic Crossan have begun to shift our spiritual consciousness into an explicit, self-aware, and intellectually rigorous eutopian mode. What we also need to grasp is that democracy can only be as humane and ecologically fruitful as the spiritual vision that undergirds democratic aspiration: spiritual vision is the humus out of which democracy grows. Borg and Crossan are makers of such compost. They are spiritual gardeners. Democracy is only superficially a set of legalized governing procedures (though those procedures are important). Democracy in its motive power arises from the liberating realization that the Earth is a gift garden we are to till in community; that, for all our superficial differences, we are a common humanity; and that the core teaching of every great spiritual tradition is peace and justice, stewardship and servanthood, Green, Rainbow, and Red.

The "antiglobalization" movement is, needless to say, not in the least opposed to a rich and ongoing cultural sharing the world over. It is fully within the eutopian spiritual tradition. It will not stop until the spears have been refashioned into pruning hooks, until the false god of utopia has been toppled, until eutopia can contemplate Earth, once again, as well-tended garden—until, as Martin Buber points out, the eutopian "Community of God" replaces an obsolescent and no longer endurable decay of nations intent on taking Earth's ecology with them into destitution and disaster.

NOTES

1. Crossan, *God*, 16–17.
2. Crossan, *God*, 152.
3. Phillips, *American*, 282.

27

Eutopian Postscript

IN "THEIR FEARFUL ENERGIES" I invoked, seemingly out of blue sky or Green Earth, a sequence of Four Ages: Mother, Father, Son, and Daughter. Let me now say how I came by such a formulation and explain what its significance might be.

I was raised on a small homestead farm in northern Wisconsin and brought up in a traditional Evangelical and Reformed Church. That particular denomination (as I recall) still adhered to a body-and-blood Eucharist, unlike many other Protestant denominations. The psychoanalytical significance is that the old Eucharist fed faith directly in the body. The bread literally became Jesus' body, the wine became his blood. The "Body of Christ" was, therefore, as in high medieval times (as the concept of "Christendom" suggests), a transhistorical organism, an evolving and growing social entity. The new symbolic Eucharist stimulated remembrance via cerebral belief. Its "Body of Christ" consisted of scattered individuals awaiting salvation at the end of time, perhaps by virtue of predestination. This underlying religious dynamic (still poorly understood in its correlation, historically, to the rise of commercialism, scientific skepticism, and utopian psychology) occurred in that period when the cosmos ceased to be perceived as an organism and became instead a machine, as Carolyn Merchant shows. Human consciousness began to be brought in from hearth fire, garden, field, and forest and was compelled to sit obediently in geometric rows for educational preparation in utopian living. Folk consciousness was no longer merely corralled and constrained by aristocratic controls; it was now beginning to be evaporated by the steady elimination of folk life, folk tradition, and folk livelihood as utopian infrastructure steadily asserted its dominance and hegemony. The eutopian body of evolving spiritual comprehension was overpowered by the utopian mind of organized supernation. Peasants and heathen were now to be "civilized."

I left both farm and church, weary of utopian economics (which grossly undervalued agrarian culture) and disgusted with utopian religion (which blindly overvalued civilized belief). But just as I discovered that the plight of small-scale agriculture wouldn't let me rest, so, too, there were religious questions that kept nagging at my mind and spirit. Who, for instance, was the Holy Ghost? Of the Three Persons in the Christian Trinity, the Holy Ghost (or Holy Spirit) defied conceptualization. Why? It was a process of painful intuition, sweated out in a cluttered storage room behind the basement furnace in a house in St. Louis (a most inauspicious "retreat center"), that led me to the realization that the Holy Ghost was feminine. I understood then why the Holy Ghost had been locked in the religious closet: God was Three Persons, and the Godhead was Male. The orthodox tradition was all too clear. As the Christian storyteller and science fiction writer C. S. Lewis had his hero Dr. Ransom say in *That Hideous Strength*: "What is above and beyond all things is so masculine that we are all feminine in relation to it."[1] *That* was the essence of the orthodox tradition, a tradition that had permeated and shaped cultural sensibility for hundreds and thousands of years. The pale Holy Ghost in the religious closet was a She, and the patriarchal officiates of Christian orthodoxy were fearful of—Her.

Having come that far—self-satisfied, perhaps—I turned to the anarchist writings of Paul Goodman, a social perspective that knitted easily with the Jeffersonian underpinnings of my youth. Goodman led to Martin Buber, Buber to Norman O. Brown, Brown to Lewis Mumford. I was, most explicitly, looking for the history of agriculture (especially the village culture of subsistence communities) over against civilization with its expropriating power. This became, simultaneously, an investigation into class, culture, gender, and religious distortion: how the (male) city was organized to expropriate commodities from the (female) countryside, *civis* lording it over the *pagus*. Mumford taught me, especially in *The City in History*, about the Neolithic revolution: gathering to horticulture to enlarging villages to cities to civilization. In the process, I gathered some of the intellectual fundamentals of feminist investigation: was there ever a "matriarchy"? Yes, said the poetic enthusiasts. No, said the more critical scholars, strong matrilocal values, to be sure, but no "matriarchy."

To some extent, the question of whether there ever was "matriarchy" began to bore me, an academic dispute. But Mumford, Bachofen, the Jungians, weird scholarly poets like Robert Graves, and even Norman Brown, kept pointing to a twilight Mother, almost a culturally infantile recollection. Matriarchy or no matriarchy, it became clear that the Mother Goddess was a historical reality—equal rights in the history of religions. Eventually, Michael Harrington's *Socialism* found its way into my cabin in northern Wisconsin. Harrington described some of the ancestors of socialist thought. He dealt, of course, with the urban Greeks and the wilderness Hebrews, with Augustine's City of God, and he paused a while with Joachim of Floris, a twelfth-century Italian monk. According to Harrington, Joachim described what he considered the three ages of human history, a schema based on the Bible and the Christian Trinity. First, Joachim said, there was the Age of the Father. That age corresponded to Judaism, the Torah, and *Old Testament*, and it was marked by monarchy, discipline, and law. Second was the Age of the Son, a period stemming from the life and death of Jesus and manifesting itself socially as love institutionalized in the church. Third, the age yet to come, an age of consecrated anarchy, the Age of the Holy Ghost.[2]

Joachim's schema seemed a valuable psychoanalytic tool, even though incomplete. It is indeed possible to identify the monotheistic religions of Judaism and Islam as the repositories of patriarchal prerogative they were and still are: social systems whose mythic core is a jealous and powerful Father. The Christian Son emerged to take on the human condition and to either (a) endure the Father's wrath for the sake of the world or (b) endure utopia's wrath for demonstrating openly the Creator's vast eutopian compassion. This age became, through Paul's and Augustine's two-edged platonic doctrine of empire-as-God's-will, love institutionalized in the church, and the secularization of this ethos produced the welfare state: love institutionalized in governmental bureaucracy. (The City of God became the City of Man.) Furthermore, both Judaism and Islam traced their religious ancestry to the patriarch Abraham, while Christianity provided a common heritage to Western "democratic" capitalism and Eastern "democratic" communism.

Despite the considerable period of time (two thousand years roughly) that separates the respective emergence of Judaism and Islam, it is possible to say that in terms of their mutual orientation toward the Father as the creator of all life and meaning, these religions are psychologically contemporaneous. They represent rival patriarchal systems. (See the story of Isaac and Ishmael in Genesis 21.) The institutionalized love of the church also found itself with two major utopian expressions: capitalism and communism. The Christian church, after its decentralized "primitive" period following the death of Jesus, became absorbed by the Roman Empire and eventually its leading expression; when Rome fell, the church remained the instrument for civilized expansion and utopian consolidation. The Protestant Reformation eventually announced the democratization of the church; the "priesthood of all believers" became both a religious and a political ideal to be achieved in human community—as led and practiced by men. As secular government took over areas of social control formerly held by the church, the impulse toward "institutionalized love" grew into secular ideology. The dominant ideology was capitalism, upwardly striving and inherently agentic. Salvation was both personal and private. But economic historians learned to see in the lower-class revolts of the Reformation (peasant revolts and Anabaptist uprisings, especially) a striving toward a more comprehensive communal ethos and identity. Industrialism and the decay of feudal society generated self-conscious economic rivals: capitalism on the one hand, all forms of collectivism on the other. Both developments claim something of a Christian heritage, and they are, indeed, secular manifestations of the Age of the Son, love institutionalized primarily in utopian form.

And then, one winter night, the configuration congealed: if it was psychoanalytically meaningful to talk about a Mother's age in the Neolithic, a Father's age in the formation of patriarchal civilization, a Son's age of "democratic" industrialism, then my storage room intuition about the Holy Ghost as feminine seemed to suggest that the modern women's movement, dovetailing with the deterioration of patriarchal hegemony and a largely cosmetic democracy, was a visible sign of a tremendous shift in gender-based religious imagery and valuation. Joachim's Age of the Holy Ghost suddenly made sense as Daughter, and the linkage of women with nature—Mother Earth, Mother Nature—made of the Daughter's age an ecological society, eutopian. The women's movement itself (Catholic nuns threatening to bolt the church unless they, too, can become priests) was prophetic of an extraordinary cultural and religious revolution.

Intellectually, one of the biggest rejoinders came from that brilliant "vernacular" priest, Ivan Illich. A big defender of decentralist and low-energy culture, Illich seemed to be saying (for instance in his *Shadow Work*) that every low-energy society known has been organized along gender lines and that "unisex" (for Illich a decidedly pejorative word) emerged only in the decadence of imperialist decay. When we get off our high-energy utopian binge, Illich seemed to say, gender will be there to greet us. The further implication seemed to be that postindustrial gender would strongly resemble preindustrial gender: women at home with the babies and men taking care of politics. This perspective, of course, contrasted sharply with Dorothy Dinnerstein's assertion in *The Mermaid and the Minotaur* that the *solution* to our gender-derived global crises would be a radically new configuration in gender roles. For Dinnerstein, men need to learn to nurture, at home, while women learn to shape public policy. But even Illich refrained from spearing the future too firmly—realizing, perhaps, that to "capture the future" is a little like going after a whale with a butterfly net. So we are left with a lot of ambiguity. Perhaps we can try to put some of these ambiguities in a meaningful sequence.

Until the Neolithic revolutions in horticulture and animal domestication, human life meant the band and the tribe, usually nomadic. Various scholars assert that as gathering enlarged into horticulture, and as growing population density made hunting less and less feasible, women achieved a rather striking spiritual and social power. Given the anxiety toward women-as-life-source that men experience, and their subsequent drive to dominate, the control of this enlarged village space, the town and early city, was appropriated politically by men. Fascination with "ideal planning" on the part of the "smothered sons" led to brutal utopia and, eventually, to religions from which the Goddess was totally excluded: patriarchal civilization in both civil execution and religious imagery. Civilizations rose and fell, enlarging their political turf and leaving monuments of ambiguous magnificence for posterity. As the Christian priesthood of all believers helped formulate a democracy of sons, the feminist movement evolved, at least in part, as an extension of the democratic principle. But for women to be equal to men *politically* forces the issue of gender imbalance *culturally,* not only in regard to contemporary institutions but also in terms of the psychoanalytical hegemony of divinity as overwhelmingly Male. And this is where we're stuck in the present period. We are becoming saturated with the *idea* of a new gender configuration, but the actual *practice* is both socially sluggish and distorted by the hegemony of utopian mythology. When equality reaches the flowering stage of spiritual *reconciliation,* we will be ready for eutopia—even as, after September 11, we are in the grip of reactionary utopianism with explicit male murderousness.

Now the Green critique of utopia—namely, that civilization is based on class exploitation, racial discrimination, resource pillage, and gender distortion—points toward an ecological society. This critique envisions an end to classes, the integration of "races," an ecological economy and lifestyle, and a fundamentally new configuration of behavior between women and men. To achieve this end, civilization must, in its utopian formation, contract in order for a eutopian social order to grow. But this is neither an "end to civilization" nor a regression to pre-Neolithic gathering and hunting. The Green vision, both practical and humane, sustains a global view while encouraging the ecological reinhabitation of the countryside, perhaps along the lines proposed by contemporary bioregionalists. But as Lewis Mumford has stated in *The Transformations of Man*, spiritual alteration "does not become operative on a collective scale until it has remoulded social institutions."[3] What are these social institutions that need remolding?

We can begin with a partial list of those utopian institutions that kill, maim, and poison: the military, the transportation system, energy-producing utilities, manufacturers of toxins, agribusiness. Let each political entity have its peacekeepers whose function it really is to keep the peace—not a euphemism for apocalyptic bullying. Let there be trains and sailing ships, bicycles and buggies, reducing to a minimum fossil fuel consumption with its by-product pollutants. Let us have solar, wind, and water energies in place of radioactive waste and acid rain. Let the manufacture of toxins be kept to a bare minimum for the sake of the Earth and living beings. And let utopian agribusiness, with all its enforced "scientific" brutality, go the way of dinosaurs, extinct. (It's not as if we weren't forewarned: Thomas Jefferson, for all the nasty contradictions in his private economic life, was a eutopian two centuries in advance of his vision. The democratic commonwealth he anticipated—a pastoral countryside, local but quality education, neighborhood self-governance, a simple industrial infrastructure of durable goods—is as applicable today as it was at the end of the eighteenth century. Moreover, we now can see how destructive the spurning of that vision has been.)

Thanks to Lewis Mumford, we now know that civilization has been an explicitly utopian project. From this basis of understanding we can go on to recognize the utopian features of those religions and secular ideologies that have allied themselves with, or been the outgrowth of, civilization. In the background, in the shadows, we can begin to perceive the outline of eutopia: the dynamics of sexual reconciliation, racial equality, cultural respect, economics within ecological tolerance. Green, Red, and Rainbow politics—colorful diversity on the living Earth—provide us with a healing path. With enough enlightened will and spiritual humility, a democratic citizenry can transform utopia into eutopia. There is time enough to effect this transformation peacefully, to undo the "permanent war economy" and utilize Earth's resources in the building of an ecological and democratic culture. But the option to peaceful change looks increasingly like breakdown and collapse, and that implies a great deal of suffering. Perhaps utopia is so firmly entrenched—both in its mythology, its ideology, and its infrastructure—that some degree of breakdown cannot be avoided. But total breakdown, as in nuclear war, is a terrifyingly real possibility, and it obviously would make moot any question of eutopian unfolding.

There seems to be no option, therefore, but to develop as best we can the spiritual discipline, the richer consciousness, and the earthier lifestyle that anticipates a eutopian order—at once a world culture and an intimate community. Yet, the truth is that we who promote ecological living in an ecological society are not ourselves practiced in eutopian skills, just as we are unpracticed in that kind and quality of intentional dialogue and spiritual humility that sustain intimate community. In the cloying distractions and deadly disasters of modern life, we are walking a very thin line between utopian extinction and eutopian transformation. "Nuclear numbness" may well be the pre-eminent symbol and symptom of utopian consciousness. When enough people begin to achieve a more sustained *eutopian* consciousness through lifestyle changes in the dimension of ecological living, racial dialogue and gender reconciliation, then Green and Rainbow politics will begin to discover important ways in which to be surprisingly and quickly effective.

These essays, written in the excitement of discovery and anxiety of distress, are a small nudge in the direction of eutopia. It is my prayer that their spiritual content is part of the healing we so desperately need.

NOTES

1. Lewis, *Hideous*, 316.
2. Harrington, *Socialism*, 18.
3. Mumford, *Transformations*, 77.

28

A Gardener's Afterword

It is somewhat unseemly to tack on additional remarks like a woodshed onto a garage onto a shack, perhaps even a homemade greenhouse (or at least a clunky cold frame) onto the south side of the woodshed. But in the aftermath of the terrorist attacks on the Twin Towers and the Pentagon on September 11, 2001, the unseemly is hardly to be shunned.

These essays—some of them—go back in earlier drafts to 1980. "Their Fearful Energies" is drawn from the concluding remarks of a sprawling, unpublished manuscript completed in '80 or '81, my first concerted attempt to grapple with and understand the meaning of the "farm crisis" in civilization and, more deeply, the consequences of civilization globally supreme.

Already I had come to a pretty firm conclusion about how the coercive inner structure of civilization (its militarism and involuntary servitude) was gender-based. This tightly organized power configuration, in contending kingdoms, fed off of primary production, with agriculture at its base, and created the basic pattern of aristocrat and peasant endemic to all civilization prior to the industrial revolution. Civilization in its industrial form has, over two centuries or slightly longer, steadily undermined or simply destroyed the coherence of rural culture, and has brought to ruin the subsistence life of the peasantry on a world scale. In its current globalization, the United States leads the effort to impose a capitalist market system on every culture worldwide. Agribusiness is inherently and thoroughly utopian. Agribusiness is what civilization does to the countryside, and to rural culture, when the opportunity arises.

The attacks on September 11 represent, in my estimation, the first major assault on the presumptions of civilization in its globalized form. I firmly believe this is far less an example of the alleged "clash of civilizations" than it is an attack on the very power centers of the New Rome, an attempt to speed up the hardening and decline of hubristic imperialism.

It is less a clash of civilizations than a clash of patriarchal monotheisms. The consequences of this "war against terrorism" (which is the prevailing language of U.S. political and military leaders) will be like those of the sorcerer's apprentice who, by chopping the broom into pieces, attempts to stop the magical process he has created: each piece, each splinter, revives as a full-blown broom, and the horror escalates into madness—until the Sorcerer intervenes, and the broom (plus Mickey Mouse, in Walt Disney's version) is restored to its original humble form.

II

Joan Chittister, a Benedictine nun, in an essay entitled "God Become Infinitely Larger" (in *God at 2000*, edited by Marcus Borg and Ross Mackenzie), says that "Augustine's hierarchy of being—maleness and a God made in a male image—has got to be at best suspect, at the least incomplete, and in the end bogus." Chittister goes on to say:

> This male construct of a male God in a male-centered world is no picture of God at all. God is not maleness magnified. God is life without end: all life, in everything, in everyone, in men and women, and in women and men. The light of the divine shines everywhere and has no gender, no single pronoun, no one image.[1]

God is surely not "maleness magnified," but the Western patriarchal religions within utopian civilization most certainly have been and still are maleness magnified. Furthermore, these religions reflect and mythologize the magnified maleness operative within civilization since its dawn; for civilization at its core is expropriation and armed control, and both behaviors depend on aggressive male energy. Male energy in possession of weapons, and with the will to use those weapons, lies at the base of the civilizing impulse. All that follows from this power impulse radiates triumphant male energy.

God is not going to intervene in any Walt Disney conclusion to this current disaster. The "Sorcerer" is not going to authoritatively set things right. This may be the war onto global exhaustion Norman O. Brown prophesied in *Love's Body*—"an end to war, an end to brotherhood." The world, on the far side of this conflict and impending carnage, will be a different world than the one we live in now.

I write these thoughts one week exactly after the terrorist disasters. The West, led (as always) by the United States, has as yet taken no military action (that I know of) against the terrorists, the Taliban, Osama bin Laden, Afghanistan, or any other organization or country on the list of suspects. No one seems to know whether more terrorist attacks are forthcoming or whether the organization has, as it were, shot its bolt. But one thing does raise its monstrous head to scan the landscape in anticipation. That one thing is a global Armageddon of retaliatory male desire. It is possible to foresee—though prudence may yet prevail—a global exhaustion of male energy in an ever-widening whirlpool of violence, a Third World War. The men who, at this moment, plan the American response to last Tuesday's attack, may be the very men whose actions pave the way for the end of utopia, the end of civilization as we know it.

Have the birth pangs of the Age of the Daughter begun in deadly earnest?

III

In the hope of someday sending this manuscript to a publisher, it's time to get out hammer, nails, a few old boards, and cobble together another lean-to or, perhaps, a homemade gallows by which to hang what little reputation may have accrued in these essays.

It is May, Mother's Day 2003, rainy and blustery, with a record number of tornados in the Midwest in the last nine or ten days. Since September 18, 2001, Afghanistan has been bombed, the Taliban destroyed, Al Qaeda dispersed, and Osama bin Laden driven into hiding. A camp for captive "illegal combatants" continues to operate at Guantanamo Bay, Cuba—a camp for which the U.S. judicial system denies any jurisdiction.[I] Iraq, first in line in the "Axis of Evil," has been inspected, bombed, invaded, and occupied. Saddam Hussein (the "Ace of Spades") has apparently fled, while about half the pack of most wanted "playing cards" have been captured or detained.[II] The U.S. is busy bringing "democracy" to Iraq by means of its vast military superiority and huge corporate power.

One element within my thesis predicts conflict to the point of exhaustion. But why should men reach exhaustion, except temporarily, when history suggests that fresh reserves of anger, rage, hate, and sheer calculated self-interest always seem in ample supply?

I. The Supreme Court did assert its jurisdiction in June 2004. But as of September 2007, this judicial assertion has not brought any relief to the captives

II. Saddam Hussein was captured, in a "spider hole," in December 2003, and hanged at the end of December 2006.

The world is becoming more unified, but less because of imposed uniformity than we usually imagine; and the women's movement is among the causes and effects of this softer transformation. It is women's energy that is the emerging force that will remove violence as the quick and easy "remedy" for conflict. The shift into dialogue and negotiation will compel, by its very nature, the recognition of legitimate grievance, and the recognition of legitimate grievance will result in the elevation of sharing and stewardship into key positions in both domestic and international policy. Once this process is entered into seriously, it will be unstoppable. The discrediting of evasion, denial, and reaction will be swift and overwhelming. The conventional utopian agenda, seemingly godlike and invincible, is about to be severely pruned and constrained.

IV

Did George W. Bush "Gore" John Kerry in November 2004? Or did Kerry (with help from the strategic leadership of the Democratic Party) "Gore" himself? Rather than follow the advice in Richard Goldstein's "Neo-Macho Man," Kerry and his handlers chose to underplay the dynamic Mommy Party elements and overplay ("John Kerry reporting for duty" with a crisp salute, "We will hunt the terrorists down and kill them," "I will conduct a smarter, more effective war in Iraq and win it") an imitation of the Daddy Party's macho swagger.

All of which suggests two options. Either macho swagger is what the American people are looking for, or, as usual, the "Left" is too timid and too reflexively deferential to boldly put forth a truly meaningful agenda for de-empiring the American military and our country's matrix of overconsuming economic systems. Both options seem true; that is, we *are* addicted to swagger and the Left (insofar as the Democratic Party actually represents the Left) *is* timid. Swagger holds sway not only because we are addicted to it, not only because of psychofundamentalist mythological rigidities involving what Marcus Borg calls "monarchical images of God," not only because we wallow in civilized fantasies of superiority, but also because the corporate funders of political campaigns (as John Nichols points out in regard to Al Gore's campaign in 2000) work to restrain candidates from articulating analyses and propounding policies that threaten corporate control. Corporations are utopian fiefdoms whose purpose is the maintenance and protection of their advantages. As such, they are inherently hostile to democratic oversight.

Therefore, we need not only a campaign finance law far more rigorous than McCain-Feingold, a law that would firmly spell out the public-only dimensions of candidacy and campaigning, but also an enlarged multiparty electoral system on the parliamentary model, so that small parties—like the Greens—can begin to get the necessary public support and political traction by which to both promote neglected and evaded issues and, in a far fairer electoral process, see those issues brought meaningfully into the legislative arena. Yet neither campaign finance reform nor proportional representation, important as they are, gets to the heart of our need for Green culture, for ecological stewardship and socialist servanthood. The prevailing governing mythologies of both religion and civilization must be surpassed precisely because those mythologies channel our consent and empower our behavior in the direction of deeper and deeper degradation and disaster. We are encased in mythologies of utopian superiority.

The question is: what is strong enough, convincing enough, culturally and ecologically gripping enough, to take the place of obsolete mythology? What can preserve our sense of the whole with tenderness and nonviolence, with reconciliation and healing? The answer seems to be: only the core ethical content of the world's religions, not the fabulous mythologies in which the ethical content is loosely embedded, but *the ethical content itself.* This core—loving God, loving Earth, and loving neighbor—must be re-embedded in a new spiritual constellation of feminine ambience universalizing an honest sense of human community and ecological integrity.

V

Meanwhile, the Christian Right has had its man in the White House. It has a lot of true believers in state legislatures and in Congress. Politically, the Christian Right has been largely running the show in this country. The Christian Right is, conceivably, the single most dangerous political force on the face of the Earth. That is so because it has congealed in itself the two most powerful forms of unreflective self-righteousness in human history. The first of these forms is civilized mythology. That is, "civilization" is only a step below God in conventional understanding. Anything "civilized" is, by definition, good—as opposed to the wickedness of that which is "primitive," "savage," "barbarian," or "pagan." Civilized consciousness (as in the crushing or extermination of Native American culture) has been unreflectively righteous. "The only good Indian is a dead Indian." Christian mythology is the other form of unreflective righteousness. Christians are the new "chosen people." Christianity is the only true religion. The Bible is the only inspired Word of God. All other religions are, finally, false and even wicked. All other religions are more than deficient. They all (with the exception of Judaism, which is only stubbornly willful in its refusal to embrace the Christian Christ and, therefore, toying with its own damnation) are also "pagan," and all true Christians know how God hates paganism.

Since the End Times are at hand, certain things need to be put in order. Israel, for instance, must expand its boundaries to "biblical" dimensions, possibly from the Nile to the Euphrates. A new temple must be built in Jerusalem, on the exact site where a very important Islamic mosque now sits. Therefore, American support for Israel, with its occupation of Palestinian territory, with its invasive settlements, and with its quick and deadly military—all this must continue and even be accelerated. There must (despite lip service to the contrary) be no "two-state" solution, much less a single, unified, and integrated state made up of Jews, Muslims, and Christians. All this helps to hasten the coming apocalypse. It is the total anticipation of the Last Days.

The thing that really unites and unifies civilized and religious mythologies is unreflective doctrinal self-righteousness. This leads to, and results in, the inability and refusal to recognize the degree and magnitude of their respective sinfulness. Unreflective righteousness blocks spiritual discernment. If contrition leads to lucidity, willful righteousness leads to mythological blindness. The respective mythologies of civilization and Christianity have been, and are, repositories for unhesitating moral certainty and ethical superiority. These mythologies have now congealed in a new global imperialism of moral, unreflective aggression.

What does this mean in terms of history? In my estimation, both civilization and Christianity (all three Abrahamic religions, actually) are truly facing their "last days." Globalization has brought the "Babylonian invention" of civilization to its logical breaking point. Civilization is in its early death throes. Especially since its unification with state power under Constantine (though its imperialistic impulses go back even farther, to the

Apostle Paul and John the Evangelist), Christianity has been absolutely convinced of its moral superiority, its singular justification before God.

What we have seen in the Bush administration, as in much of the Republican leadership, is this convicted, unreflective moral superiority at work, at the controls of the mightiest military the world has ever known. The "last days" really are at hand, only the outcome is not at all what the Christian Right anticipates or expects.

VI

Given the magnitude of the office and the incredible power that goes with it, it is tempting to call George W. Bush's presidency a kind of heroic destiny: the perfect man for a moment of millennial crisis. But given the mediocre simplicity of the man, with his wheedling voice of aggressive petulance—no particular aptitude for art or history, not a curious traveler, no deep lover of nature, a religious "believer" of the most depressingly literalistic sort—one wonders what gave him this most ill-fitting sense of confidence, a relentless pushing for policies favorable to the military-industrial complex, corporate America embellished with endless red, white, and blue bunting.

There seem to have been two sources to this confidence, one of which is certain and the other ambiguous. The uncertain source (because one never knows how much is political posturing) is conventional, fundamentalist, religious mythology. This is the triumphalist imperialism of a form of Christian doctrine that asserts itself as totally and unconditionally true, both its view of the nature of reality and its anticipation of how the future will unfold according to certain interpretations of specific Christian scriptures. Being on the right side of God's Plan apparently makes one close to invincible, a rather firm form of confidence. The other source of Bush's confidence (as I write, in early February 2005, Bush's White House lawyer, Alberto Gonzales, who wrote the memos justifying torture of "terrorists" and "illegal combatants," and Condoleezza Rice, former National Security Advisor who ignored intelligence warnings on the impending attacks of September 11 and who worked mightily in behalf of the fictional "mushroom cloud" fear-mongering leading up to Gulf War II, have been confirmed, respectively, as Attorney General and Secretary of State) is more obvious and mundane. It is the sense of entitlement that is heavily a part of upper-class, neoaristocratic identity. It is the unreflective mantle of privilege. It is the presumption of right and authority, and it blends perfectly with Christian triumphalism and American Exceptionalism.[III] It comes with being born into a family of the rich and powerful, very much used to calculating its way and achieving its calculation. It is the prerogative of advantage.

George Bush has been the chosen, groomed spokesman and public figurehead for those who wished to maximize and, to the fullest possible extent, institutionalize the policies of neoconservative corporate advantage. He has had the immeasurable weight of that advantage behind him, guiding and praising. It has to have been a heady experience for a mediocre man. And, insofar as George W. Bush recognizes his mediocrity, his intellectual ordinariness, he may well fall back on divine election as the premise of his ascendancy, so that all these factors meld into a sort of cosmic mandate that liberates Bush's mind (and the process whereby he has selected his advisors and cabinet officers) from any hesitation, uncertainty, and ambiguity.

The perfect man for a moment of millennial crisis.

VII

Jonathan Schell, in the August 14/21, 2006, issue of *The Nation*, in a long essay entitled "Too Late for Empire," argues that a combination of nuclear and conventional weapons have made empire in the Roman or British mode unattainable for would-be world powers.[2] Kevin Phillips, in his *American Theocracy*, says that

> If the God of Tim LaHaye and the *Left Behind* series turns out to be otherwise preoccupied—as he seems to have been during the later wartime embroilments of Spain, Holland, and Britain—then, lacking heavenly intervention, any major military confrontation in the Middle East or anywhere else could easily have results similar to those of the past. U.S. global supremacy could drain away more in five to twenty years than most Americans would have thought possible.[3]

(Phillips' subsection, "War: The Military and Economic Unmaking of Global Hegemons," pages 339–344, is a concentrated gem I highly recommend.)

The utopian system has reached its limit. Civilized weapons are so toxic that their full use would eradicate mammalian life on Earth. The fossil energy used to empower the utopian economy now rebounds in cli-

III. See especially "Secular and Religious Sectarianism" in Chapter 2 of Seymour Lipset's *American Exceptionalism: A Double-Edged Sword.*

mate changing "blowback." Rampant urbanization strengthened by uninhibited population growth in conjunction with the systemic destruction of rural and agrarian livelihoods has created huge festering urban sores on Earth's body. Toxic chemicals from industrial production now show up in polar bears at the North Pole. Glaciers everywhere are melting.

The utopian system is the radiant energy of traditional, aggressive, male-dominated civilization. Its continuation (not to speak of its expansion) is possible only with intensified ecological and cultural havoc. It is a great and sobering thing to realize that the awesomely destructive conflagration depicted in Revelation is now the leering, lustful *desire* of a huge core constituency of the Republican Party. We might say that these people are the hardcore suicide cult within the utopian impulse: as utopia speeds toward the wailing wall, the Armageddon juggernaut folks grow as manic as Dr. Strangelove. They are rapture bombers on a global mission.

The thing is, in comparison to the "wartime embroilments of Spain, Holland, and Britain" that Phillips elucidates in his book, the United States not only has apocalyptic weaponry in its arsenal, it also has a very organized and influential End Times constituency so apparently convinced of its righteousness that provoking Armageddon is, for them, nothing less than a consummate act of holiness. They consider themselves the faithful welcoming committee for a radiant, vengeful Christ, who will slay unbelief and unbelievers. So while Jonathan Schell may be right—it's too late for empire—Kevin Phillips may also be right—U.S. global supremacy could dissipate more in five to twenty years than most of us would ever have thought possible. As Phillips goes on to say:

> We have seen the unfortunate precedents. Militant Catholicism helped undo the Roman and Spanish empires; the Calvinist fundamentalism of the Dutch Reformed Church helped to block any eighteenth-century Dutch renewal; and the interplay of imperialism and evangelicalism led pre-1914 Britain into a bloodbath and global decline. The possibility that something similar could propel the United States into war in the Middle East—and that once again, God would decline to rescue his chosen people—is the precedent that needs to be kept in mind.[4]

Christianity has become the sadistic Antichrist burrowed within the utopian predilections of civilized arrogance. We will only get out of or through this epochal mess by profound spiritual transformation. Repentance is the clean way, the way with minimal violence or structural wreckage. Universal (or near-universal) male exhaustion from war is another way through, a horrifically brutal and destructive way, but the way the world will go if repentance is (as I fear it will be) contemptuously dismissed as weak, puerile, and disgusting.

VIII

The alternatives are increasingly obvious. In a world of clear ecological limits to utopian expectation, civilized desire and religious belief in sanctified hallucination, we will either (unless we blow Earth to hell) bypass democratic self-governance into the restoration of aristocratic rule (even as the process of aristocratic restoration is heavily disguised as patriotic and even "democratic," with war and terror alerts its unrelenting chess moves) or we will openly confront the aristocratic and utopian elements within our cultural evolution, elements that have never been purged from our democratic experiment precisely because they were and are so deeply embedded in our civilized preconceptions and religious mythology.

Democracy must of necessity be primarily Earth-based and eutopian. It must be elementally humane and ecologically prudent. The women's movement, the rise of female energy, is what will decide the question of utopian aristocracy versus eutopian democracy. And this, in turn, hinges on the extent to which female sensibility is (or is not) more deeply grounded than male sensibility in servanthood and stewardship, in sharing and caring. I am betting that it is. The excesses of male conflict agitate the rise of female energy. The way is paved with rubble.

Utopia's days are numbered. Consuming less, sharing more, we can create a beautiful global culture of eutopian proportion and ecological balance. Utopia's long reign is ending. The first brightenings of the eutopian dawn are apparent to the early riser. The gardener is in the garden.

NOTES

1. Chittister, *God*, 68.
2. Schell, "Late," 13–24.
3. Phillips, *American*, 339.
4. Phillips, *American*, 375.

Bibliography

Abbey, Edward. *Down the River*. New York: Dutton, 1982.

Acworth, Evelyn. *The New Matriarchy*. London: V. Gollancz, 1965.

Altizer, Thomas J. J. *The Gospel of Christian Atheism*. Philadelphia: Westminster, 1966.

Aries, Philippe. *Centuries of Childhood*. Translated by Robert Baldick. New York: Knopf, 1962.

Asimov, Isaac. *Foundation*. New York: Avon, 1951.

———. *Foundation and Empire*. New York: Avon, 1952.

———. *Second Foundation*. New York: Avon, 1953.

Badian, E. *Roman Imperialism in the Late Republic*. Ithaca, NY: Cornell University Press, 1971.

Bakan, David. *The Duality of Human Existence: An Essay on Psychology and Religion*. Chicago: Rand McNally, 1966.

Beard, Mary. *On Understanding Women*. New York: Greenwood Press, 1968.

Bellamy, Edward. *Looking Backward*. New York: Dutton, 1960.

Berry, Wendell. "Back to the Land: The Radical Case for Local Economy." *The Amicus Journal* (Winter 1999).

———. *The Gift of Good Land*. San Francisco: North Point Press, 1981.

Borg, Marcus. *The God We Never Knew*. San Francisco: Harper, 1997.

Boulding, Elise. "Peace Culture: The Vision and the Journey." *Friends Journal* (September 2005).

———.*The Underside of History: A View of Women through Time*. Newbury Park, CA: Sage, 1992.

Braverman, Harry. *Labor and Monopoly Capital: The Degradation of Work in the Twentieth Century*. New York: Monthly Review Press, 1998.

Brown, Norman O. *Life Against Death: The Psychoanalytical Meaning of History*. Middletown, CT: Wesleyan University Press, 1959.

———. *Love's Body*. New York: Vintage, 1966.

Brutus [pseudo.]. *Confessions of a Stockbroker: A Wall Street Diary*. New York: Bantam, 1973.

Buber, Martin. *Paths in Utopia*. New York: Macmilllan, 1950.

———. *Two Types of Faith: A Study of the Interpenetration of Judaism and Christianity*. Translated by Norman P. Goldhawk. New York: Harper, 1961.

Burbach, Roger, and Patricia Flynn. *Agribusiness in the Americas*. New York: Monthly Review Press, 1980.

Burke, Carol. "Why They Love to Hate Her." *The Nation*, March 22, 2004.

Caro, Robert. *Master of the Senate*. New York: Knopf, 2002.

Carroll, James. *Constantine's Sword: The Church and the Jews*. Boston: Houghton Mifflin, 2001.

———. *House of War: The Pentagon and the Disastrous Rise of American Power*. Boston: Houghton Mifflin, 2006.

Childe, V. Gordon. *What Happened in History*. Baltimore: Penguin, 1954.

Chittister, Joan. "God Become Infinitely Larger." In *God at 2000*, edited by Marcus Borg and Ross Mackenzie. Harrisburg, PA: Morehouse, 2000.

Claremont de Castillo, Irene. *Knowing Woman: A Feminine Psychology*. New York: Putnam, 1973.

Clough, Shepard B. *The Rise and Fall of Civilization*. New York: Columbia University Press, 1961.

Columbia Encyclopedia. 2nd ed. Morningside Heights, NY: Columbia University Press, 1950.

Commoner, Barry. *The Closing Circle: Nature, Man, and Technology*. New York: Knopf, 1971.

Crankshaw, Edward. *Khrushchev's Russia*. Baltimore: Penguin, 1959.

Crossan, John Dominic. *God and Empire: Jesus against Rome, Then and Now*. New York: HarperCollins, 2007.

Dale, Tom, and Vernon Gill Carter. *Topsoil and Civilization*. Norman, OK: University of Oklahoma Press, 1955.

Dinnerstein, Dorothy. *The Mermaid and the Minotaur*. New York: Harper & Row, 1976.

Dyer, Joel. *Harvest of Rage: Why Oklahoma City is Only the Beginning*. Boulder, CO: Westview Press, 1997.

Edwards, Mike. "Ukraine." *National Geographic*, May 1987.

Ehrenreich, Barbara. *Blood Rites*. New York: Metropolitan, 1997.

Epstein, Barbara. "What Happened to the Women's Movement?" *Monthly Review*, May 2001.

Fischer, Ernst. *The Necessity of Art: a Marxist Approach*. Baltimore: Penguin, 1963.

Frank, Thomas. *What's the Matter with Kansas? How Conservatives Won the Heart of America*. New York: Metropolitan, 2004.

Freud, Sigmund. *Civilization and Its Discontents*. Translated by James Strachey. New York: W. W. Norton, 1961.

Galbraith, John Kenneth. *The New Industrial State*. Boston: Houghton Mifflin, 1967.

Galeano, Eduardo. "Terror in Disguise." *The Progressive*, February 2003.

Gilk, Paul. *Nature's Unruly Mob: Farming and the Crisis in Rural Culture*. Millville, MN: Anvil Press, 1986.

Goldstein, Richard. "Butching up for Victory." *The Nation*, January 26, 2004.

———. "Cartoon Wars." *The Nation*, February 21, 2005.

———. "Neo-Macho Man: Pop Culture and Post-9/11 Politics." *The Nation*, March 24, 2003.

Goodman, Paul. *Compulsory Miseducation*. New York: Horizon, 1964.

Goodwyn, Lawrence. *The Populist Moment: A Short History of the Agrarian Revolt*. New York: Oxford University Press, 1978.

Guthman, Julie. *Agrarian Dreams*. Berkeley: University of California Press, 2004.

Haldeman, H. R. *The Ends of Power*. New York: Times Books, 1978.

———. *The Haldeman Diaries: Inside the Nixon White House*. New York: G.P. Putnam's, 1994.

Harrington, Michael. *Socialism*. New York: Saturday Review Press, 1970.

———. *Toward a Democratic Left*. New York: Macmillan, 1968.

Hauser, Arnold. *The Philosophy of Art History*. New York: Knopf, 1959.

Hedges, Chris. *American Fascists: The Christian Right and the War on America*. New York: Free Press, 2006.
Heilbroner, Robert. *The Great Ascent: The Struggle for Economic Development in Our Time*. New York: Harper and Row, 1963.
———. *The Worldly Philosophers: The Lives, Times, and Ideas of the Great Economic Thinkers*. New York: Simon and Schuster, 1961.
Hobsbawm, E. J. *Industry and Empire*. Baltimore: Penguin, 1968.
Illich, Ivan. *Deschooling Society*. New York: Harper & Row, 1971.
———. *Shadow Work*. Boston: Marion Boyars, 1981.
Jackson, Jesse, Jr. "George Bush's Democrats." *The Nation,* January 22, 2001.
Kingsolver, Barbara. "A Good Farmer." *The Nation,* November 3, 2003.
Kropotkin, Peter. *Fields, Factories, Workshops*. New York: Harper & Row, 1975.
Kunstler, James Howard. *The Long Emergency: Surviving the Converging Crises of the Twenty-First Century*. New York: Atlantic Monthly Press, 2005.
Lakoff, George. *Don't Think of an Elephant!* White River Junction, VT: Chelsea Green Publishing Company, 2004.
Lefabvre, Georges. *The Coming of the French Revolution*. Translated by R. R. Palmer. Princeton, NJ: Princeton University Press, 1947.
Leopold, Aldo. *A Sand County Almanac*. New York: Oxford University Press, 1966.
Lerner, Gerda. *The Creation of Patriarchy*. New York: Oxford University Press, 1986.
Lewis, C. S. *The Great Divorce*. New York: Macmillan, 1946.
———. *The Last Battle*. New York: Macmillan, 1956.
———. *That Hideous Strength*. New York: Macmillan, 1968.
Lievan, Anatol. "Liberal Hawk Down." *The Nation,* October 25, 2004.
Lipset, Seymour. *American Exceptionalism: A Double-Edged Sword*. New York: Norton, 1996.
Mailer, Norman. *The Prisoner of Sex*. Boston: Little, Brown, 1971.
Manning, Richard. *Against the Grain: How Agriculture Hijacked Civilization*. New York: North Point Press, 2004.
Marcuse, Herbert. *An Essay on Liberation*. Boston: Beacon, 1969.
———. *Eros and Civilization: A Philosophical Inquiry into Freud*. Boston: Beacon, 1955.
Marx, Leo. *The Machine in the Garden: Technology and the Pastoral Ideal*. New York: Oxford University Press, 1964.
Melman, Seymour. *The Permanent War Economy: American Capitalism in Decline*. New York: Simon and Schuster, 1974.
Merchant, Carolyn. *The Death of Nature: Women, Ecology and the Scientific Revolution*. New York: Harper & Row, 1980.
Meredith, Martin. *Nelson Mandela: A Biography*. New York: St. Martin's Press, 1998.
Merton, Thomas. *Gandhi on Non-Violence: A Selection from the Writings of Mahatma Gandhi*. New York: New Directions, 1964.
Morris, William. *News from Nowhere*. New York: Viking Penguin, 1984.
Mumford, Lewis. *The City in History*. New York: Harcourt, Brace & Jovanovich, 1961.
———. *The Myth of the Machine*. New York: Harcourt, Brace & World, 1966.
———. *The Pentagon of Power*. New York: Harcourt, Brace & Jovanovich, 1970.
———. *The Story of Utopias*. New York: Viking, 1962.
———. *The Transformations of Man*. New York: Harper & Row, 1956.
———. "Utopia, the City, and the Machine." In *Interpretations and Forecasts: 1922–1972*. New York: Harcourt, Brace & Jovanovich, 1979.

Murphy, Cullen. *Are We Rome? The Fall of an Empire and the Fate of America.* Boston: Houghton Mifflin, 2007.

The New Testament. Nashville, TN: The Gideons International, 1973.

Nichols, John. Review of *The Next Agenda: Blueprint for a New Progressive Movement*, edited by Robert Borosage and Roger Hickey. *The Progressive,* March 2001.

Philllips, Kevin. *American Theocracy: The Peril and Politics of Radical Religion, Oil, and Borrowed Money in the 21st Century.* New York: Viking, 2006.

Polanyi, Karl. *The Great Transformation.* Boston: Beacon, 1957.

Priest, Dana. *The Mission: Waging War and Keeping Peace with America's Military.* New York: W. W. Norton & Company, 2003.

Rauschenbusch, Walter. *Walter Rauschenbusch: Selected Writings,* edited by Winthrop S. Hudson. Mahwah, NJ: Paulist Press, 1984.

Rifkin, Jeremy, and Ted Howard. *The Emerging Order.* New York: Putnam, 1979.

Roszak, Theodore. "Introduction." In *Small is Beautiful: Economics as if People Mattered.* New York: Harper & Row, 1973.

Roy, Arundhati. *War Talk.* Cambridge, MA: South End Press, 2003.

Ruether, Rosemary Radford. *New Woman/New Earth.* New York: Seabury Press, 1975.

Ruskin, John. *Unto This Last.* Lincoln, NE: University of Nebraska Press, 1967.

Salisbury, Harrison E. *Russia.* New York: Atheneum, 1965.

Salm, Peter. *The Poem as Plant: A Biological View of Goethe's Faust.* Cleveland: Press of Case Western Reserve University, 1971.

Schell, Jonathan. "The New Nuclear Danger." *The Nation,* June 25, 2001.

———. "Too Late for Empire." *The Nation,* August 14, 2006.

Schumacher, E. F. *Small is Beautiful: Economics as if People Mattered.* New York: Harper & Row, 1973.

Smith, Henry Nash. *Virgin Land: The American West as Symbol and Myth.* New York: Vintage, 1950.

Snow, C. P. *The Two Cultures and the Scientific Revolution.* Cambridge: Cambridge University Press, 1993.

Stone, Merlin. *When God Was a Woman.* New York: Harcourt, Brace & Jovanovich, 1976.

Tawney, R. H. *The Acquisitive Society.* New York: Harcourt, Brace and Howe, 1920.

Taylor, Charles. *The Ethics of Authenticity.* Cambridge, MA: Harvard University Press, 1992.

Thompson, William Irwin. *The Time Falling Bodies Take to Light: Mythology, Sexuality, and the Origins of Culture.* New York, St. Martin's Press, 1981.

Tocqueville, Alexis de. *Democracy in America.* New York: Knopf, 1945.

Toynbee, Arnold. *The Industrial Revolution.* Boston: Beacon, 1957.

Toynbee, Arnold J. *Civilization on Trial.* New York: Oxford University Press, 1948.

Unger, Craig. *House of Bush, House of Saud: The Secret Relationship between the World's Two Most Powerful Dynasties.* New York: Scribner, 2004.

Veblen, Thorstein, *The Theory of the Leisure Class.* Boston: Houghton Mifflin, 1973.

Webster's New International Dictionary. 2nd ed. Springfield, MA: G&C Merriam, 1910.

Weisskopf, Walter. *The Psychology of Economics.* Chicago: University of Chicago Press, 1955.

Williams, Patricia. "Power and the Word." *The Nation,* February 28, 2005.